Feminist Ecologies

Lara Stevens • Peta Tait • Denise Varney
Editors

Feminist Ecologies

Changing Environments in the Anthropocene

palgrave
macmillan

Editors
Lara Stevens
University of Melbourne
Melbourne, VIC, Australia

Peta Tait
LaTrobe University
Melbourne, VIC, Australia

Denise Varney
University of Melbourne
Melbourne, VIC, Australia

ISBN 978-3-319-64384-7 ISBN 978-3-319-64385-4 (eBook)
https://doi.org/10.1007/978-3-319-64385-4

Library of Congress Control Number: 2017950845

Cover illustration: r e a, PolesApart series 2009, triptych no.2/a, 100 x 92 cm, c-type photograph, edition of 10

Printed on acid-free paper

This Palgrave Macmillan imprint is published by Springer Nature
The registered company is Springer International Publishing AG
The registered company address is: Gewerbestrasse 11, 6330 Cham, Switzerland

Acknowledgements

Thanks are due to the celebrated scholars Freya Mathews, Kate Rigby, Deborah Bird Rose and Ariel Salleh for the opportunity to republish their ecofeminist essays and to their original publishers: *Environmental Ethics, the Trumpeter, Arena Journal*, and the Native Title Research Unit. We are grateful to Christine Bryan at the Australian National University Archives Program and Noel Butlin Archives Centre for her assistance with a previously unpublished paper by the late Val Plumwood.

We would like to express our sincere thanks to Jill Orr and r e a and collaborators for permission to include images of their performances. We are grateful to Deb Hart and Liz Conor for their generous sharing of the Climate Guardians' experiences and to Maggie Miles for allowing us to reprint her photograph of the Climate Guardians at the Eiffel Tower in 2015.

Lara is very grateful for the support of Gill Stevens, Peter Stevens, Estelle Paterson, Lachie Paterson and Eddie Paterson. Peta sincerely thanks Annie McGuigan for her ongoing love and support, and she gives a big thank you to the wonderful members of the New Ecological Discourses reading group. Denise thanks Vic Marles, CEO of the Trust for Nature, an extraordinarily progressive organization, which holds the annual International Women's Day Breakfast in Melbourne that brings feminism and environmentalism together in such a dynamic and inspiring way. Many thanks also to friend and prominent Australian environmental scientist Geoff Westcott from Deakin University for conversations about the science of climate change. Finally we thank the Australia Centre at the University of Melbourne, especially Ken Gelder and Amanda Morris, for unstinting support in the long-term development and outcomes of this project.

The articles and papers in Part I: Foundational Ecofeminisms were originally published as:

Salleh, Ariel. 1984. Deeper Than Deep Ecology: The Eco-Feminist Connection. *Environmental Ethics* 6 (Winter): 339–345.

Mathews, Freya. 1994. Relating to Nature: Deep Ecology or Ecofeminism? *The Trumpeter: Journal of Ecosophy* 11 (4) (Fall): 159–166.

Rigby, Kate. 1998. Women and Nature Revisited: Ecofeminist Reconfigurations of an Old Association, *Arena Journal* 12: 143–169.

Rose, Deborah Bird. 1995. Women and Land Claims. In *Land, Rights, Laws: Issues of Native Title*, 1–8. Canberra: Native Title Research Unit, Australian Institute of Aboriginal and Torres Strait Islander Studies.

Plumwood, Val. 2003. Ecofeminist Analysis and the Culture of Ecological Denial. Unpublished conference paper from the World Life-Culture Forum: The Transformation of the 21st Century and Life-Culture Sallim held in Suwon, South Korea. Made available by the Butlin Archives, Australian National University Library.

Contents

1

Introduction: 'Street-Fighters and Philosophers': Traversing Ecofeminisms

Lara Stevens, Peta Tait, and Denise Varney

Feminist Ecologies: Changing Environments in the Anthropocene emerges at the intersection of two progressive twentieth-century political movements: one concerned with the fight for women's rights and the other with ecological sustainability within the environment. The book celebrates the ongoing philosophical and activist advocacy of feminist ecologies as it traces the ecofeminist movement's roots and alignment with recent social, cultural and artistic developments. It proposes the broad term 'feminist ecologies' to capture the diversity of the movement over the last 45 years and the range of possible ways in which feminist and ecological concerns can speak to one another in the era of the Anthropocene.

The capacity of feminist ecologies to reveal the interconnectedness of environmental and social injustices makes it an urgent and timely field of inquiry for the current moment. This collection arises out of the need to address the challenges of climate change, land degradation, species

L. Stevens (✉) • D. Varney
The University of Melbourne, Melbourne, VIC, Australia

P. Tait
La Trobe University, Melbourne, VIC, Australia

L. Stevens et al. (eds.), *Feminist Ecologies*,
https://doi.org/10.1007/978-3-319-64385-4_1

extinction and the disproportionate effects of these changes upon particular communities of women and their livelihoods. *Feminist Ecologies* shows why ecofeminist thought, as it has become known, remains pertinent today by demonstrating how its key actions, writings and thinking underpin twenty-first-century feminist perspectives on ecological debates. The book demonstrates the progressive development of this thinking and its activism.

This collection grounds its historical moment in the Anthropocene. Atmospheric scientist Paul J. Crutzen theorized the Anthropocene as a new geological epoch shaped by the actions of humankind (2002: 23). The concept has been taken up and responded to by numerous environmental philosophers, humanities scholars as well as activists and it provides a useful starting point for the ecofeminist critiques that appear in *Feminist Ecologies*. It neatly evokes the contradiction between the human causes of environmental destruction and the human capacity to protect and care for the biosphere. For Crutzen, the Anthropocene marks a historical period that can be linked to the atmospheric effects of the industrial revolution after the eighteenth century.[1] The expansion of capitalism with the industrial revolution connects: technological advancement; the attendant change in political economy; the realization of many of the ambitions of Enlightenment science; and the enabling of unprecedented exploitation of the natural world. Accordingly ecofeminism critiques global capitalism, because of its accompanying exploitation of the 'others': women, the poor, the colonized and the nonhuman.

But even with this advancement in naming the Anthropocene, the inherent social inequity of such drastic and rapid environmental change is not illuminated and the idea of the Anthropocene might even imply that all humanity is equally responsible. *Feminist Ecologies* reminds readers of the established field of ecofeminist knowledge: about how environmental injustice is linked to social injustice, particularly gendered social injustice. The overarching question of the book is: how do contemporary feminist and ecological scholarship, activism and artistic practice progress ecofeminist thinking in response to environmental problems?

Ecofeminism, like feminism, has always been concerned with challenging and changing the oppressive structures that imbue the lives of women and men. The predominance of female scholars in this field and in this book is a historical legacy. Feminist thought is vital if we are to redress the near universal neglect of women by cultures around the world, including academic ones. This book shows how feminism and ecofeminism continue to evolve in social practice.[2]

Despite the escalating challenges of climate and rapid environmental change, and the persistence of widespread abuse and exploitation of women all over the world, ecofeminism is rarely discussed in public debate and is overlooked in much recent academic discourse. From the early 1980s, subjects on ecofeminism were offered in Western universities. Yet it cannot be taken for granted what exactly ecofeminism is or does, either historically or today. This book offers a range of perspectives on what ecofeminism is, does or can do. For example, Ariel Salleh calls ecofeminism 'a strategy for social action', stressing that it is not a static ontological claim about 'the nature of women' (1993: 231). Freya Mathews writes that: 'ecofeminism is by no means a position or a theory, but simply a wide open field of enquiry' (1994: 62). Misconceptions about ecofeminism also pose challenges to its varied movements as, like many women-led causes, it sometimes has to fight to be recognized and treated seriously by other environmental, political and social activists. As Kate Rigby notes, ecofeminism has to work hard to show it is more than 'simply a naïve form of feminine nature worship' (1998: 168).

By the 1990s there were multiple feminisms, which accordingly impacted ecofeminism and expanded it in new and varied directions. As with most feminist movements or waves, there remains no singular and agreed upon definition of ecofeminism. Carolyn Merchant lists some of the different types of ecofeminism: liberal ecofeminism, radical ecofeminism, cultural ecofeminism, social ecofeminism, socialist ecofeminism, ecological ecofeminism, deep-ecology ecofeminism, transformative ecofeminism, aboriginal ecofeminism and developing world ecofeminism (1996: 207). This book follows in this interdisciplinary tradition with contributors from a broad range of scholarly backgrounds—from philosophy to theatre and performance studies, political economy to gender studies, history to religious studies. These diverse perspectives mean that it addresses the junctures between masculinist political and cultural attitudes and behaviours, and their effects on the natural world in a variety of contexts.

Ecofeminism continues to be important because of its interdisciplinarity and therefore its usefulness in addressing the complexity of environmental and social crises today. As Salleh notes:

> Ecofeminism is the only political framework I know of that can spell out the historical links between neoliberal capital, militarism, corporate science, worker alienation, domestic violence, reproductive technologies, sex tourism, child molestation, neocolonialism, Islamophobia, extractivism, nuclear weapons, industrial toxics, land and water grabs, deforestation, genetic engineering, climate change and the myth of modern progress. (2014: ix)

The chapters in this collection address pertinent public questions around misogyny, gender equality, justice, democratic ethics, environmental protection and sustainability. The ecofeminism coming out of the northern hemisphere today is particularly focused on exploring the silences around race, species and sexuality. *Feminist Ecologies* recognizes these areas and directly considers the question of race with regards to Indigeneity and Indigenous women. There is, however, more work to be done in investigating the work of southern hemisphere ecofeminists and their engagement with issues of animal and plant life as well as human sexuality in the era of the posthuman (Braidotti 2013).

In the wider context of this field, *Feminist Ecologies* raises a number of questions: why have women environmental activists and thinkers played such a pioneering role in expanding ecological thought? And how has their work influenced global ecological and feminist activist movements? What is productive about bringing the terms feminism and ecology together? Our collection does not exhaustively or definitively answer these questions. Instead, it considers the dialectical relationship between environmental and feminist causes; the relational identities of feminists and ecofeminists; the possibility of framing ecofeminism as another wave of feminism gaining momentum; and the value of thinking about identity politics, activist histories and feminist movements over time. In exploring these points of tension, *Feminist Ecologies* helps situate contemporary ecofeminism as a complex, controversial, layered, varied and multidisciplinary project.

Feminist Ecological Theory

The volume looks to the future by recalling the momentum of the past. It brings pioneering texts, written in the 1980s and 1990s by preeminent ecofeminist thinkers, into dialogue with new research in the 2010s. The book is organized into two parts.[3] Part I: Foundational Ecofeminisms, includes four previously published articles, and one unpublished paper by key scholars in the field. Part II: Ecofeminist Currents, consists of eight chapters of recent ecofeminist scholarship that draw on the foundational texts as well as contemporary philosophy, and ecocritical and feminist thought. It charts a history of women's recent activism in order to contextualize it, and to reveal its legacies and continuities.

The theoretical advances in this book challenge the dominant historical ways that humans have related to, conceptualized, sympathized and interacted with the planet and biosphere. Its unique combination of historical and contemporary work on the intersections between gender and ecology has significant implications for global environmental concerns. *Feminist Ecologies* sets out to better understand present and future challenges for feminism and the environment by situating them in relation to earlier ecofeminist writings. As such it maps an ongoing dialogue between women's issues and rapid ecological change within the Anthropocene.

Much of the best-known ecofeminist scholarship emerged in the early 1990s with key texts such as Val Plumwood's *Feminism and the Mastery of Nature* (1993); Greta Gaard's *Ecofeminism: Women, Animals, Nature* (1993a); Maria Mies and Vandana Shiva's *Ecofeminism* 1993/2014; Carol J. Adams' *The Sexual Politics of Meat* (1992); Carolyn Merchant's *Earthcare: Women and the Environment* (1996); as well as important ecofeminist collections such as Karen Warren and Barbara Wells-Howe's *Ecological Feminism* (1994). As Gaard summarizes, ecofeminist approaches consider 'the fundamental interconnectedness of all life' and resist all forms of oppression including the oppression of nonhuman species (1993b: 2–3).

Early ecofeminist thinking was interdisciplinary and it developed its unique thought through a diverse range of scholarly fields including analytic and continental philosophy (e.g., Mathews 1991; Plumwood 1993); political economy and sociology (e.g., Eckersley 1992; Salleh 1997); anthropology, ethnography, ethics, law and Indigenous land rights (e.g., Rose

1992, 1996; Moreton-Robinson 2013); religious studies and poetics (e.g., Mews and Rigby 1999); and cultural, literary and performance studies (e.g., Adams 1992; Chaudhuri 1994). New volumes and essays expand on this interdisciplinary vein in cultural and performance studies (e.g., Arons and May 2012; Gaard et al. 2013; Varney 2015; Stevens 2015; Tait 2015), and social ethics, history and animal studies (e.g., Tait 2012; Chaudhuri 2017).

The editors of *Feminist Ecologies* employ a transhistorical methodology to locate the significant (but also under-circulated and under-recognized) work of key ecofeminist thinkers. The writers who appear in 'Foundational Ecofeminisms' have made long-standing and significant contributions to ecological and ecofeminist thought throughout their careers. Pioneering ecofeminist scholars and activists—Ariel Salleh, Freya Mathews, Kate Rigby, Deborah Bird Rose and Val Plumwood—are recognized and celebrated internationally for their contribution to ecofeminist scholarship. At the forefront of an ecofeminist movement that expanded from the 1970s, their work, both then and now, informs and underpins debates about environmental adaptation and sustainability as well as intersecting debates about women's rights, and race and class inequalities under the conditions of global capitalism.

Plumwood's groundbreaking book *Feminism and the Mastery of Nature* (1993) earned her a place within the male-dominated international canon of environmental philosophers. Rose's *Dingo Makes Us Human: Life and Land in an Australian Aboriginal Culture* (1992) won the Stanner Prize for a work on Aboriginal issues and became so popular that it is currently in its third edition. Salleh consolidated her comprehensive thinking in the landmark book *Ecofeminism as Politics: Nature, Marx and the Postmodern* (1997/2017), which brought together the shared interests of ecology, feminism, socialist and colonial resistance. Mathews is a leading figure in ecological philosophy (eco-philosophy or ecosophy), persistently arguing for a different approach to the way humans perceive and position themselves within the biosphere by connecting with Aboriginal Australian philosophy. She is particularly well known for her Spinoza-inspired work on panpsychism that first appeared in *The Ecological Self* (1991). As a world leader in the study of ecopoetics and environmental studies in the humanities, Rigby has made a significant contribution to advancing understanding of the links between ecologies and spirituality, particularly in *Ecology, Gender and the Sacred* (1999), co-edited with Constant Mews. We have chosen to represent a piece of each of these writers' early work to demonstrate their originality and foresightedness, and to show the influence and progressive development of their radical thinking.

The emergent field of the Environmental Humanities replicates multi-disciplinary ecofeminism as it too encompasses the fields of anthropology, history, philosophy, cultural studies, religious studies, animal studies, Indigenous studies, poetry, law and social and ecological justice. For the women whose work appears in 'Foundational Ecofeminisms', ecofeminism is not simply an abstract theoretical project or scholarly exercise. Rather, it is firmly integrated into their ways of living. Salleh notes that '[e]cological feminists are both street-fighters and philosophers' (2014: ix) and this is true for all the pioneering thinkers represented in *Feminist Ecologies*.[4] The women featured in this collection are also active on behalf of environmental causes and practice the ecofeminist theory they champion through their political advocacy for feminist causes.

Glocal Australia

In keeping with ecofeminist principles, *Feminist Ecologies* demonstrates how the global relies on the local, that is, on place. Most of the scholars and activists whose work appears in the book were born and raised in Australia and New Zealand. Rose, who was born in the United States, has spent much of her life living and working in Australia, particularly in remote Aboriginal communities. The editors were concerned that the contributions to this volume belong to a particular place but its ecologies reflect both the local and the global. The effort of communities to enact sustainable practices and/or protect threatened ecosystems is happening at a grass-roots level across the globe in the twenty-first century. Paradoxically, giving attention to sustainable practices at a local level might be what constitutes the shared experience of humanity in the twenty-first century. Whether it involves maintaining a garden for food, involvement in an anti-pollution campaign, or marching to raise awareness about rape, the most powerful global responses to ecofeminist issues today are happening at a local level—at the level of the glocal. The politics of the glocal is foundational to ecofeminist thinking and is made more powerful by the expansion of social media networks and the speed of information sharing in the twenty-first century via the Internet and mobile phone capacities.

The capacity of national governments to implement policy and regulation for ecological sustainability varies enormously. This is nowhere more apparent than in the United Nations' efforts to grapple with divergent political difference in the struggle to reach an agreement on carbon reduction to slow the global rates of climate change. At the local level, however, individual households and communities everywhere implementing sustainable practices and this common approach is strongly women-centred in the area of food production. People who are faced with environmental problems that threaten their survival are mobilizing powerful local movements, often led by women from impoverished communities, fighting large corporate interests and political inertia (see Klein (2015) for examples of where this grass-roots activism is effective). The local versus the global power imbalance is an ongoing feature of ecofeminist analysis in the Anthropocene because women and children are particularly vulnerable.

The decision to centre this collection on Australian ecofeminists builds on and enlarges existing international perspectives (e.g., Gaard et al. 2013). The Australian focus, however, comes at the expense of an in-depth engagement with the extraordinary ecofeminist work that has come out of, for example, India, and parts of Asia, Africa, South America, North America and Europe over the last half century. To put together a truly representative international collection is a project that deserves attention, but it is beyond the scope of this book. The editors hope that *Feminist Ecologies* will set a precedent for further volumes coming out of these other particular glocal environments.

In the meantime, there is much to be learnt from the Australian focus. Merchant writes that 'women in Australia are providing leadership in reversing ecological damage and in developing an ecofeminist ethic of earthcare' and notes that in the 1990s they were 'leading an ecological revolution' (1996: 186). The vibrancy of the field in Australia and the kinds of work coming out of this nation prompt the editors to ask: what is unique about the Australian context that has inspired such prolific and pioneering work in this field? The kinds of research presented in Ecofeminist Currents suggest that Australia's colonial history combined with its high agricultural and resource extractive capitalist economy, which has caused severe environmental degradation since settlement in 1788, might go some way to explaining the strong ecofeminist response.

Modern Australia has active industries in coal and other mining, forestry, sheep and cattle farming, which have led to issues of soil and sand erosion, high salinity levels, over-use of toxic pesticides polluting the soil and water as well as declining air quality. Yet Australia's Aboriginal and Torres Strait Islander people have historically viewed the land as having intrinsic value and thus have tended to practise an ethic of care and respect towards the flora and fauna, as well as the land's seasons and ecological systems. While colonization in Australia has meant that Aboriginal Australians are rarely given a voice, particularly one that might threaten to contradict capitalist profiteering, environmentalists and especially ecofeminists have demonstrated a far greater interest in Aboriginal knowledge and care for 'country'. Aboriginal philosophies, stories and specialized knowledge of the Australian land have been of great interest to the foundational ecofeminists featured in this collection, who have listened, adopted and advocated these ideas, and woven them into their unique forms of ecofeminist thought and ecoactivism. In some local Australian contexts, such as Martu country in the Western Desert and the Mardoowarra River in the Kimberley, where Anne Poelina advocates for 'traditional ecological knowledge', Aboriginal ideas of guardianship over country are already being put into practice.[5]

Despite Australia's legacy of environmental activism, innovative thinking in deep ecology and formative ecofeminism it has, more recently, experienced the rise of conservative governments with strong electoral mandates that have resisted and overturned action on policies promoting sustainability and greenhouse gas reduction. While the background to this book is recognition of activism in one national context, political battles in Australia are indicative of what has occurred in other developed countries. Social conservatism aligned with neoliberalism's prioritization of corporate interests has fuelled regressive strategies that intensify the struggle in all areas of activism and reinforce the need for strong theoretical analysis.

Despite the collection's emphasis on Australian ecofeminist thinkers, the local Australian environmental and feminist movements have always been strongly global. All the ecofeminist scholars in 'Foundational Ecofeminisms' have worked in universities in Europe, America and Australia and have been advocates for environmental and feminist causes that have implications well beyond the borders of their homeland. This collection's representation of important Australian ecofeminist thinkers is, however, far from exhaustive and there are many important Australian ecofeminists, some of whom worked collaboratively with Salleh, Mathews, Rigby, Rose and Plumwood.[6]

The local-global nexus that figures strongly in the collection is particularly important given the intersecting nature of climate and/or rapid environmental change and feminist concerns. The effects of environmental change often have their strongest impact on people at a local level but, as we know, global consensus and commitment to change is required to mitigate problems such as global warming.

Ecofeminism and New Materialisms

There has been a significant increase in scholarship that addresses ecological questions over the last decade. The relatively new field of the Environmental Humanities is growing in universities in the United States, the United Kingdom, Europe and Australia. As the Environmental Humanities and the broader fields of ecology and sustainability gain traction around the world, this collection offers a timely contribution in its consideration of how ecofeminism informs and expands on the debates that are central to these multidisciplinary areas of research.

One important strand in the Environmental Humanities is what has become known as 'new materialisms' with its connected theories of 'agential realism' (Barad 2007), 'vibrant matter' (Bennett 2010) and 'object-oriented-ontology' (Harman 2002), to name a few. The works of new materialist thinkers demand attention be paid not only to the interconnectedness of the human and nonhuman but also to the agency of nonliving matter and its effects on living organisms (Coole and Frost 2010).[7] For many ecofeminists, such views of the natural world are not necessarily 'new'. As Merchant wrote in 1996, ecofeminists see that 'nature is an active subject, not a passive object to be dominated' (in Rigby 1998: 155). The challenge of women's agency, which has always been at the heart of feminist advocacy, is complicated by the 'agency' of nonhuman life, even seemingly inert or non-living matter. But the risk that human inequities might become displaced encourages an ecofeminist interpretation of new materialisms.

In recognizing the 'material turn' in the humanities, this collection also notes the fundamental tension between ecological politics and historical feminism and a contradiction in bringing these terms together under the name 'feminist ecologies'. Ecology demands that we pay attention to nonhuman matter, what phenomenological approaches call the 'more-than-human' world (Abram 1996), and situate the human within a wider web of interrelations. Yet, feminist movements have historically confronted the hierarchies of power; recognized different levels of privilege and agency afforded to women of different ages, races, sexual orientation and ability; and attended to the rights and freedoms of women. Feminism has thus necessarily been anthropocentric in orientation and some feminists might reject the repositioning of women within a new set of interrelations that is so attuned to the nonhuman. It might be seen as merely another excuse for women's needs and desires to be subsumed, this time under the greater good of a healthy biosphere, where women are left to clean up the mess of scientific rationalism and the industrial revolution.

This raises a potential tension between historic feminist goals and those of ecofeminism. Bearing this tension in mind, we nevertheless argue that the frictions between feminism's anthropocentrism (or as Mathews describes, its individualism) and ecofeminism's attentiveness to the nonhuman might also provide productive sites for critically rethinking patriarchal relations and attitudes to women and the natural world. It achieves this by showing how different forms of patriarchy not only determine hierarchical social structures but also, necessarily, instrumentalize and exploit the nonhuman world. It demonstrates how patriarchal attitudes towards the woman-nature nexus have been historically intertwined and the variety of ways in which these associations and stereotypes persist in contemporary attitudes and practices.

Rigby notes that 'a reactionary reinscription of the woman-nature connection is forever hovering menacingly over the ecofeminist project' (1998: 145). She sees ecofeminist scholars who want to affirm women's difference as a model of alternative knowledge and practices as performing a 'tricky tightrope act' between essentialism and poststructuralism's views on difference (Rigby 1998: 167). By asking if ecofeminism is feminist, Victoria Davion finds that the fight against patriarchy and the fight to save the environment are 'inextricably interconnected' at a conceptual level (Davion 1994: 11). Some ecofeminists fear talk of embodiment as a form of biological determinism that feminists have worked so long and hard to resist. Yet for many ecofeminists the conceptual and the embodied experience of the woman-nature connection are equally indivisible. They see humans as nature-in-embodied-form and use women's historically different experiences to deconstruct the man over woman, humanity over nature, mind over body, production over reproduction dualisms of Ancient Greek philosophy, scientific rationalism, or contemporary capitalist patriarchalism.

Salleh as a sociologist researches the conditions of the global majority of women engaged in reproductive labour and is among a number of ecofeminist scholars who view women as privileged agents of ecorevolution because culturally and socially they have been placed in closer proximity to the natural world. Their epistemic advantage does not derive from their biology, yet their actions of nurturing in the contexts of childrearing, as much as environmental advocacy, provide potential models for a non-instrumental ethic of care towards the biosphere. As Salleh writes:

> Women are certainly embracing ecological responsibility, so much so, that it has even been remarked that it looks like they are being used all over again in their traditional housekeeping role as unpaid keepers of oikos at large. (1993: 237)

Salleh sees that sustainability cannot be adequately addressed until the relations between production and reproduction are rearranged between men, women and nature (1993: 239). Her work has emphasized that women worldwide labour as an often invisible class who can skillfully integrate head, hand, heart, and womb to grow life and protect the conditions of life. While women traditionally have been forced to labour on terms defined by men and in institutions controlled by men, she offers examples of labours that generate unique ecological knowledges for dealing with natural relations and processes. She describes this kind of knowing as 'an embodied materialism'—a concept that reaches beyond a 'greener future with gender equity' and even beyond eco-socialism.

Changing Environments

Part: Foundational Ecofeminisms, opens with a republication of Ariel Salleh's major early work, 'Deeper Than Deep Ecology: The Eco-Feminist Connection' (1984), an article which became widely known and sparked heated academic debate over the next decade. In it, Salleh questions whether deep ecology is a 'sociologically coherent position' and attacks the gendered blind spots of the work of two of the founders of the movement; Arne Naess and Bill Devall. In particular, she is critical of deep ecology's attempt to replace anthropocentrism with ecocentrism in such a way that it ends up glossing over the link between the exploitation of nature and the exploitation of women. As such, she accuses the deep ecologist position of being 'politically and historically static' (341). She is further troubled by the deep ecologist's advocacy for birth control programmes which she describes as 'another grab at women's special potency' (340).

In 'Relating to Nature: Ecofeminism and Deep Ecology' (1994), renowned philosopher Freya Mathews demonstrates her thoughtful way of probing contradictions and reconciling philosophical complexities with great clarity for the reader. This piece offers an important analysis of the intersections between movements in ecofeminism and deep ecology. She offers a philosophical account of the problems of deep ecology which advocates for the human and nonhuman merging into 'undifferentiated oneness' (162). She writes that '[e]cofeminists, in contrast, tend to portray the natural world as a community of beings, related, in the manner of a family, but nevertheless distinct' (159). Despite this difference, she shows how both movements share a desire to break down the binary thinking that has set nature in opposition to culture and resist the hostility to nature that is built into patriarchal ideology. Mathews argues that the human impetus to 'save' the world might in fact come out of our human capacity to destroy it.

In 'Women and Nature Revisited: Ecofeminist Reconfigurations Of An Old Association' (1998), Kate Rigby offers a comprehensive survey of ecofeminism to show how it became a well-established interdisciplinary field of scholarship by the end of the twentieth century. Rigby succinctly and carefully evaluates and compares major publications in the mid-1990s, particularly the work of Carolyn Merchant, Mary Mellor and Ariel Salleh. She explores how these ecofeminists engage theoretically with marxism, socialism, and social ecology and their shared critique of the capitalist patriarchal enforcement of women's role as 'reproductive labour'. She shows how ecofeminist thinking offers a variety of possible approaches to address our present ecological predicament. Rigby's accompanying critique of spiritualist thinking and its dilemmas remain particularly pertinent.

Deborah Bird Rose writes in 'Women and Land Claims' (1995) about the gender bias in the uptake of land rights by Aboriginal Australians following the Mabo High Court decision in 1992, and the passing of the Native Title Act in 1993. These legal landmark cases gave Aboriginal Australians a chance to exercise rights over the land for the first time since colonization, albeit within a Western legal system. Rose explains that this historic moment, in which Aboriginal people could take advantage of economic, political and social opportunity, was often denied to Aboriginal women. This was due to a legal system in the early 1990s in Australia, which was dominated by male judges, legal counsel and anthropologists and thus privileged male Indigenous plaintiffs. Rose argues that, particularly in the Northern Territory land claims, Aboriginal women were often not granted the right to speak their knowledge and status as land owners. She also investigates exceptions to this trend where judges made special arrangements for Aboriginal women to reveal secret/sacred knowledge as evidence under conditions that were acceptable to those women. The ongoing effects of this gender bias continue to affect Aboriginal Australian women today as Ambelin Kwaymullina and Maryse Helbert show in the 'Feminist Currents' section of the book.

The final chapter in 'Foundational Ecofeminisms' presents an unpublished paper, 'Ecofeminist Analysis and the Culture of Ecological Denial' (2003), by Val Plumwood, one of the world's best-known and respected ecofeminist scholars. In her writing at the beginning of the twenty-first century, Plumwood expanded her philosophical critique to examine the failures of liberal democracy, in particular its separation of the public and private spheres, its hostility to collective action, its privileging of upper income groups, its racism and its masculinist conception of citizenship, which discounts the private sphere and thus women's labour. In this chapter, Plumwood emphasizes philosophical thinking as she elaborates on the 'hyper-separation' of human and nature that sees environmental degradation as a matter of overwhelming concern for the long-term sustainability of life. She argues that the separation of the human and the nonhuman leads to the denial of ecological life, an act that perpetuates environmental destruction.

Part II of the book, Ecofeminist Currents, provides some of the most recent scholarship in ecofeminism. These chapters document a history of recent activism as they explore environmental and women's movements and look to contemporary examples of ecofeminist thinking, writing and creating. Chapter 7, 'From *The Female Eunuch* to *White Beech*: Germaine Greer and Ecological Feminism' by Lara Stevens reveals how an ecological consciousness has been intrinsic to the thinking of globally renowned feminists such as Germaine Greer. Connecting present and past feminisms, Stevens rereads Greer's 1970s book *The Female Eunuch* to find it was not only a radical and groundbreaking exposé of patriarchal oppression of women, but its systemic critique can be better understood through the subsequent work of ecofeminists Plumwood, Mathews, Rigby, Salleh and Rose. Further, she reveals how Greer's proto-ecofeminist thinking is developed and expanded in *White Beech* (2013), a book which describes Greer's current project of protecting and rejuvenating 60 hectares of Australian rainforest.

Denise Varney's 'Climate Guardian Angels: Feminist Ecology and the Activist Tradition' considers the Australian activist ensemble, the Climate Guardians, whose all female members dress as angels in long white gowns adorned with large swooping organza wings. She describes how these secular modern angels appear, perform visitations, gather and manifest en masse in public spaces, including in Paris for COP 21 in November 2015. She argues that this fascinating ensemble draws on angel iconography, not to re-activate tradition, but to dissolve the divisions between the human and nonhuman world, to reject dualisms even as they evoke them and to appropriate and radicalize spectacular iconography to contest inaction on climate change in the public arena.

In Chapter 9, 'Thinking–Feminism–Place: Situating the 1980s Australian Women's Peace Camps', Alison Bartlett links ecology and epistemology in what she terms 'epistecology'. She considers how situated knowledge is a legacy of 1980s feminist and peace activism. Drawing specifically on the women's peace movement at Greenham Common in the United Kingdom and at Pine Gap in Australia, which protested war, masculinity, militarism and its continuum with violence against women and environmental destruction, she argues they both reveal the potency of place and embodiment, and were foundational to ecofeminist activism.

'Performing Ghosts, Emotion and Sensory Environments' by Peta Tait draws together ideas of ecofeminism and eco-phenomenology as it asks: what can performance and art that reflects feminist values with an ecological emphasis contribute to a political understanding of the human and the nonhuman? The performance artist Jill Orr, and Aboriginal Australian multimedia artist r e a, create performances that present transhistorical gender ambiguous figures framed by dystopic environments. These works evoke the ecofeminist concerns of forced migration, colonial occupation, climate change and the loss of land for human and nonhuman species. Tait discerns that the bodily responses and emotional disturbances that such artworks provoke in human spectators encourage phenomenological comprehension of the impact of bodies within shared environments.

In Chapter 11, 'You are on Indigenous Land: Ecofeminism, Indigenous Peoples and Land Justice', Ambelin Kwaymullina insightfully explains how the environments that are contested in mining, forestry, climate change and other environmental accords and negotiations are those to which Indigenous peoples belong. She explains that the relationships of Indigenous women to their homelands are grounded in narratives that practise a way of knowing that pre-dates Western conceptions of feminism and ecofeminism. In this sense, Indigenous women's knowledge is a rich source of land-based epistemologies, ontologies and axiologies. But the engagement of Indigenous ways of knowing, being and doing with those of the West is fraught by colonial complications, including the long-standing privileging of Western ways of knowing and exploitative research practices that have appropriated the knowledge and denied the power of Indigenous women. In this context, shifts within Western knowledge disciplines to more holistic and inclusive ways of knowing—such as ecofeminism—present both a challenge and an opportunity.

Anne Elvey's 'Feminist Ecologies in Religious Interpretation: Australian Influences' charts the recent history of feminist interpretation of biblical religion and the evolving, but uneven, relationship between feminist and ecological thinking in biblical studies. While it describes how ecological feminism did not appear explicitly in biblical interpretation until the 1990s with the Earth Bible project in Australia, and it remains marginalized, the chapter explores theological scholar Elaine Wainwright's writings for their important contribution. Elvey argues that Wainwright's organizational leadership and publications, encompassing the postcolonial contexts of Australia, New Zealand and the Pacific region, reveal complex intersections of religion, race and ecological thought.

In Chapter 13, 'Australian Women in Mining: Still a Harsh Reality', Maryse Helbert argues that Australia's economic growth, which has been heavily dependent on its capacity to dig and extract natural resources for the world market, produces environmental and social injustice. In this chapter, Helbert uses an ecofeminist approach to connect different oppressive structures in order to understand the uneven distribution of the costs and benefits for men and women in mining projects and towns such as the Bowen Basin, the Cooper Basin and the Western Cape York regions in Queensland. She proposes that an ecofeminist ethics can help locate alternatives to correct the unequal gendered distribution of the risks and benefits of mining projects in Australian communities.

'In the interest of all mankind': Women and the Environmental Protection of Antarctica' by Emma Shortis explores the role of women activists in the successful World Park Antarctica campaign of the 1980s, which culminated in one of the most significant international environmental agreements in existence today: the indefinite ban on Antarctic mining and the comprehensive environmental protection of an entire continent. The very few studies of the history and development of this agreement almost invariably privilege the role of diplomats and political leaders who were (and still are) almost overwhelmingly white men. Shortis argues that knowledge of the central role of women in Antarctic history and politics within the broader field of feminist and gender-based approaches to environmental protection is crucial to understanding the past, present and future of the continent.

As environmental issues increasingly demand urgent attention and gender inequalities persist in the twenty-first century, this collection offers a timely examination of how ecofeminist thought, its histories and its activism intersect with global debates about environmental catastrophe as a product of global corporate capitalism and neoliberalism. Further, it redresses the frequent erasure of ecofeminist scholars from the contemporary ecological field as it seeks to give gender prominence in the environmental and social challenges of today. To find solutions to ecological and feminist issues we need new modes of theory and praxis, activism and philosophizing as well as radical rethinking of policy, law, spirituality and education. *Feminist Ecologies* sets us on this path. It challenges us to take control over the Anthropocene and shift our environments towards new and more sustainable directions.

Notes

1. The Anthropocene is a controversial periodizing term even without the inclusion of its gender implications (Rose 2008; Baskin 2015; Haraway 2015). Disagreements ensued over whether the new geological moment of the Anthropocene should commence with the beginning of agriculture or the early nineteenth-century industrial age or around 1950, when atomic bombs left radioactive traces on the earth's surface. Despite this, the term Anthropocene has been widely adopted in the Environmental Humanities where it is usually considered to begin with the industrial age.
2. The language of feminism and ecofeminism also changes over time. For example, in some contexts in the 1980s, the word 'men' was sometimes used but this was predominantly replaced by 'patriarchy' by the 1990s in order to emphasize the way structural forces manifest in the lives of everyone.
3. The impetus and ideas for this book came out of the 'Feminist Ecologies' conference held at the University of Melbourne on 13–14 November 2014; Germaine Greer gave the keynote lecture ironically titled 'Mother? Nature?' in which she spoke about her current land conservation project at Cave Creek rainforest in Queensland.
4. Plumwood was well known in environmental and social activist circles throughout her life, particularly for her struggle to save Australia's old growth forests as well as for her advocacy for Aboriginal women's rights, abortion rights and refugee rights. Salleh has a long personal history of grass-roots activism, including being active in the Movement Against Uranium Mining, the Franklin Dam Blockade, the Australian Greens Party, the Women in Science Enquiry Network, the Society for Social Responsibility in Engineering, the Australian Government's Gene Technology Ethics Committee and the International Sociological Association Research Committee for Environment & Society. Mathews has a long-time involvement in land restoration and currently

manages a biodiversity reserve on a rocky outcrop in semi-arid northern Victoria. Together with two new co-owners, she has recently established a private Conservation Trust to protect the property in perpetuity. Rose lived for many years in remote Indigenous communities in the Northern Territory of Australia and has served as a consultant anthropologist for the Aboriginal Land Commissioner, Northern Land Council, Central Land Council, NSW Parks and Wildlife Service and the Aboriginal Areas Protection Authority. In her attendance at science-dominated international gatherings on the environment, Rigby has championed the importance of humanities scholarship. Her work questions the way that fundamental Christian beliefs support the human dominance of nature and has been a prominent figure in the development of ecotheology.

5. Dr Anne Poelina is a Nyikina Traditional Custodian of the Fitzroy River in the West Kimberley region of Western Australia. http://majala.com.au/our-people/Custodian. See also the Martu Living Deserts project which combines modern science with Indigenous ecological knowledge in a partnership between the Martu people of the Western Desert, the Nature Conservancy and BHP Billiton http://www.natureaustralia.org.au/our-work/lands/martu-living-deserts/. (Accessed 21/5/2017).
6. These include, for example, Annette Greenall Gough, Jo Vallentine, Janis Birkeland, Patsy Hallen and numerous others.
7. This might include Manuel DeLanda, Rosi Braidotti, Timothy Morton, Bruno Latour, Diana Coole, Samantha Frost and Richard Grusin.

Bibliography

Abram, David. 1996. *The Spell of the Sensuous*. New York: Vintage Books.

Adams, Carol J. 1992. *The Sexual Politics of Meat: A Feminist-Vegetarian Critical Theory*. New York, NY: Continuum.

Arons, Wendy, and Theresa J. May, eds. 2012. *Readings in Performance and Ecology*. Basingstoke: Palgrave Macmillan.

Barad, Karen. 2007. *Meeting the Universe Halfway: Quantum Physics and the Entanglement of Matter and Meaning*. Durham, NC: Duke University Press.

Baskin, Jeremy. 2015. Paradigm Dressed as Epoch: The Ideology of the Anthropocene. *Environmental Values* 24: 9–29.

Bennett, Jane. 2010. *Vibrant Matter*. Durham: Duke University Press.

Braidotti, Rosi. 2013. *The Posthuman*. Hoboken: Wiley.
Chaudhuri, Una. 1994. 'There Must Be a Lot of Fish in that Lake': Toward an Ecological Theater. *Theater* 25 (1): 23–31.
———. 2017. *The Stage Lives of Animals*. London: Routledge.
Coole, Diana, and Samantha Frost, eds. 2010. *New Materialisms: Ontology, Agency and Politics*. Durham and London: Duke University Press.
Crutzen, Paul J. 2002. Geology of Mankind: The Anthropocene. *Nature* 415 (6867): 23.
Davion, Victoria. 1994. Is Ecofeminism Feminist? In *Ecological Feminism*, ed. Karen Warren with Barbara Wells-Howe, 8–28. London: Routledge.
Eckersely, Robyn. 1992. *Environmentalism and Political Theory: Toward an Ecocentric Approach*. Albany: State University of New York Press.
Gaard, Greta, ed. 1993a. *Ecofeminism: Women, Animals, Nature*. Philadelphia: Temple University Press.
———. 1993b. Living Interconnections with Animals and Nature. In *Ecofeminism: Women, Animals, Nature*, ed. Greta Gaard, 1–12. Philadelphia: Temple University Press.
Gaard, Greta, Simon C. Estok, and Serpil Oppermann, eds. 2013. *International Perspectives in Feminist Ecocriticism*. London: Taylor and Francis.
Haraway, Donna. 2015. Anthropocene: Capitalocene, Plantationocene, Chthulucene: Making Kin. *Environmental Humanities* 6: 159–165.
Harman, Graham. 2002. *Tool-Being: Heidegger and the Metaphysics of Objects*. Peru, IL: Open Court.
Klein, Naomi. 2015. *This Changes Everything*. London: Penguin.
Mathews, Freya. 1991. *The Ecological Self*. London and New York: Routledge.
———. 1994. Relating to Nature: Deep Ecology or Ecofeminism? *Trumpeter* 11 (4): 159–166.
Merchant, Carolyn. 1996. *Earthcare: Women and the Environment*. New York: Routledge.
Mews, Constant J., and Kate Rigby, eds. 1999. *Ecology, Gender and the Sacred*. Clayton, VIC: Centre for Studies in Religion and Theology, Monash University.
Mies, Maria, and Vandana Shiva. 2014 [1993]. *Ecofeminism*. London: Zed Books.
Moreton-Robinson, Aileen. 2013. Towards and Australian Indigenous Women's Standpoint Theory. *Australian Feminist Studies* 28 (78): 331–347.
Plumwood, Val. 1993. *Feminism and the Mastery of Nature*. London and New York: Routledge.

Rigby, Kate. 1998. Women and Nature Revisited: Ecofeminist Reconfigurations of an Old Association. *Arena Journal* 12: 143–169.
Rose, Deborah Bird. 1992. *Dingo Makes Us Human: Life and Land in an Australian Aboriginal Culture*. Cambridge: Cambridge University Press.
———. 1996. *Nourishing Terrains: Australian Aboriginal views of Landscape and Wilderness*. Canberra: Australian Heritage Commission.
———. 2008. Love in the Time of Extinctions. *Australian Journal of Anthropology* 19 (1): 81–84.
Salleh, Ariel. 1993. Class, Race and Gender Discourse in the Ecofeminism/Deep Ecology Debate. *Environmental Ethics* 15 (3): 225–244.
———. 1997/2017. *Ecofeminism as Politics: Nature, Marx, and the Postmodern*. London and New York: Zed Books.
Stevens, Lara. 2015. Seeing the Forest for the White Beech Tree: Germaine Greer and Ecofeminism. *Plumwood Mountain, An Australian Journal of Ecopoetry and Ecopoetics* 2 (1). Accessed June 1, 2017. https://plumwood-mountain.com/seeing-the-forest-for-the-white-beech-tree-germaine-greer-and-ecofeminism-by-lara-stevens/
Tait, Peta. 2012. *Wild and Dangerous Performances: Animals, Emotions, Circus*. London: Palgrave Macmillan.
———. 2015. Love, Fear and Climate Change: Emotions in Drama and Performance. *Publication of the Modern Languages Association of America (PMLA)* 130 (5): 1501–1505.
Varney, Denise. 2015. 'Beauty Tigress Queen': Staging the Thylacine in a Theatre of Species. *JASAL, Association for the Study of Australian Literature* 15 (2): 1–13.
Warren, Karen with Barbara Wells-Howe, eds. 1994. *Ecological Feminism*. London: Routledge.

Lara Stevens holds a PhD in Theatre and Performance Studies from the University of Melbourne. She is the author of *Anti-War Theatre After Brecht: Dialectical Aesthetics in the Twenty-First Century* (Palgrave Macmillan 2016), and editor and translator of *Politics, Ethics and Performance: Hélène Cixous and the Théâtre du Soleil* (Re.press 2016), essays by French feminist philosopher Hélène Cixous. In 2014 she was the Hugh Williamson Postdoctoral Fellow in the Australian Centre at the University of Melbourne.

Peta Tait is a professor at La Trobe University, a Fellow of the Australian Academy of the Humanities and a visiting Professorial Fellow at the University of Wollongong. She has written 60 scholarly articles, chapters, 10 plays and performance texts, and her recent books include *Fighting Nature: Travelling Menageries, Animal Acts and War Shows* (2016); the co-edited *The Routledge Circus Studies Reader* (2016), *Wild and Dangerous Performances* (Palgrave Macmillan 2012), *Circus Bodies* (2005), and *Performing Emotions* (2002). She is editor of *Great Stage Directors* volume one: *Stanislavsky, Antoine and St Denis* (Bloomsbury under contract) and author of 'Emotion' (Bloomsbury under contract). Her most recent play, *Eleanor and Mary Alice*, about Eleanor Roosevelt meeting with Mary Alice Evatt and human rights, art and war, was restaged in Perth in 2016.

Denise Varney is Professor of Theatre Studies in the School of Culture and Communication at the University of Melbourne, and co-director of the Australian Centre. She was co-convenor of the Feminist Research Working Group (IFTR) from 2011 to 2015. She publishes on Australian Theatre, Brechtian and contemporary German theatre, feminist criticism and performance. Her latest book is *Performance Feminism and Affect in Neoliberal Times* (Palgrave Macmillan 2017), co-edited with Elin Diamond and Candice Amich. She is co-author of *Theatre in the Asia Pacific: Regional Modernities in the Global Era* (Palgrave Macmillan 2013), and is currently working on an Australian Research Council project, 'Patrick White and Australian Theatrical Modernism'.

Part I

Foundational Ecofeminisms

2

Deeper than Deep Ecology: The Eco-Feminist Connection

Ariel Salleh

beyond that perception of otherness lies the perception of psyche, polity and cosmos, as metaphors of one another.
John Rodman (1977)

In what sense is eco-feminism 'deeper than deep ecology?' Or is this a facile and arrogant claim? To try to answer this question is to engage in a critique of a critique, for deep ecology itself is already an attempt to transcend the shortsighted instrumental pragmatism of the resource-management approach to the environmental crisis. It argues for a new metaphysics and an ethic based on the recognition of the intrinsic worth of the nonhuman world. It abandons the hardheaded scientific approach to reality in favor of a more spiritual consciousness. It asks for voluntary simplicity in living and a nonexploitive steady state economy. The

Originally published as: Salleh, Ariel. 1984. Deeper Than Deep Ecology: The Eco-Feminist Connection. *Environmental Ethics* 6 (Winter): 339–345.

A. Salleh (✉)
University of Sydney, Sydney, NSW, Australia

L. Stevens et al. (eds.), *Feminist Ecologies*,
https://doi.org/10.1007/978-3-319-64385-4_2

appropriateness of these attitudes as expressed in Naess' and Devall's seminal papers on the deep ecology movement is indisputable (1973, 1980). But what is the organic basis of this paradigm shift? Where are Naess and Devall 'coming from,' as they say? Is deep ecology a sociologically coherent position?

The first feature of the deep ecology paradigm introduced by Naess is replacement of the Man/Nature dualism with a *relational total-field image*, where man is not simply 'in' his environment, but essentially 'of' it. The deep ecologists do not appear to recognize the primal source of this destructive dualism, however, or the deeply ingrained motivational complexes which grow out of it (See Salleh 1981, 1983, 1985). Their formulation uses the generic term *Man* in a case where use of a general term is not applicable. Women's monthly fertility cycle, the tiring symbiosis of pregnancy, the wrench of childbirth, and the pleasure of suckling an infant, these things already ground women's consciousness in the knowledge of being coterminous with Nature. However tacit or unconscious this identity may be for many women, bruised by derogatory patriarchal attitudes to motherhood, including modern male-identified feminist ones, it is nevertheless 'a fact of life.' The deep ecology movement, by using the generic term *Man*, simultaneously presupposes the difference between the sexes in an uncritical way, and yet overlooks the significance of this difference. It overlooks the point that if women's lived experiences were recognized as meaningful and were given legitimation in our culture, it could provide an immediate 'living' social basis for the alternative consciousness which the deep ecologist is trying to formulate and introduce as an abstract ethical construct. Women already, to borrow Devall's turn of phrase, 'flow with the system of nature.'

The second deep ecology premise, according to Naess, is a move away from anthropocentrism, a move toward *biological egalitarianism* among all living species. This assumption, however, is already canceled in part by the implicit contradiction contained in Naess' first premise. The master-slave role which marks man's relation with nature is replicated in man's relation with woman. A self-consistent biological egalitarianism cannot be arrived at unless men become open to both facets of this same urge to dominate and use. As Naess rightly, though still somewhat anthropocentrically, points out, the denial of dependence on Mother/Nature and the compensatory drive to mastery which stems from it have only served to alienate man from his true self. Yet the means by which Naess would realize this goal of species equality is through artificial limitation of the human population. Now putting the merits of Naess' 'ends' aside for the moment, as a 'means' this kind of intervention in life processes is supremely rationalist and technicist, and quite at odds with the restoration of life-affirming values that is so fundamental to the ethic of deep ecology. It is also a solution that interestingly enough cuts right back into the nub of male dependence on women as mothers and creators of life—another grab at women's special potency, inadvertent though it may be.

The third domain assumption of deep ecology is the *principle of diversity and symbiosis*: an attitude of live and let live, a beneficial mutual coexistence among living forms. For humans the principle favors cultural pluralism, an appreciation of the rich traditions emerging from Africa, China, the Australian Aboriginal way, and so on. These departures from anthropocentrism, and from ethnocentrism, are only partial, however, if the ecologist continues to ignore the cultural inventiveness of that other half of the human race, women; or if the ecologist unwittingly concurs in those practices which impede women's full participation in his own culture. The annihilation of seals and whales and the military and commercial genocide of tribal peoples are unforgivable human acts, but the annihilation of women's identity and creativity by patriarchal culture continues as a fact of daily existence. The embrace of progressive attitudes toward nature does little in itself to change this.

Deep ecology is an *anti-class posture*; it rejects the exploitation of some by others, of nature by man, and of man by man, this being destructive to the realization of human potentials. However, sexual oppression and the social differentiation that this produces are not mentioned by Naess. Women again appear to be subsumed by the general category. Obviously the feminist ecological analysis is not 'in principle' incompatible with the anti-class posture of deep ecology. Its reservation is that in bypassing the parallel between the original exploitation of nature as object-and-commodity resource and of nurturant woman as object-and-commodity resource, the ecologist's anti-class stance remains only superficially descriptive, politically and historically static. It loses its genuinely deep structural critical edge. On the question of political praxis though, there is certainly no quarrel between the two positions. Devall's advocacy of loose activist networks, his tactics of nonviolent contestation, are cases in point (Salleh 1984b, c). Deep ecology and feminism see change as gradual and piecemeal; the violence of revolution imposed by those who claim 'to know' upon those who 'do not know' is an anathema to both.

The fight against *pollution and resource depletion* is, of course, a fundamental environmental concern. And it behooves the careful activist to see that measures taken to protect resources do not have hidden or long-term environmental costs which outweigh their usefulness. As Naess observes, such costs may increase class inequalities. In this context he also comments on the 'after hours' environmentalist syndrome frequently exhibited by middle-class professionals. Devall, too, criticizes what he calls 'the bourgeois liberal reformist elements' in the movement—Odum, Brower, and Lovins, who are the butt of this remark. A further comment that might be made in this context, however, is that women, as keepers of *oikos*, are in a good position to put a round-the-clock ecological consciousness into practice. Excluded as many still are from full participation in the social-occupational structure, they are less often compromised by the material and status rewards which may silence the activist professional. True, the forces of capitalism have targeted women at home as consumer par excellence, but this potential can just as well be turned against the systematic waste of industrialism. The historical significance of the domestic labor force in moves to recycle, boycott, and so on has been grossly underestimated by ecologists.

At another level of analysis entirely, but again on the issue of pollution, the objectivist attitude of most ecological writing and the tacit mind-body dualism which shapes this means that its comprehension of 'pollution' is framed exclusively in external material terms. The feminist consciousness, however, is equally concerned to eradicate ideological pollution, which centuries of patriarchal conditioning have subjected us all to, women and men. Men, who may derive rather more ego gratification from the patriarchal status quo than women, are on the whole less motivated to change this system than women are. But radical women's consciousness-raising groups are continually engaging in an intensely reflexive political process, one that works on the psychological contamination produced by the culture of domination and helps women to build new and confident selves. As a foundation for social and political change, this work of women is a very thorough preparation indeed.

The sixth premise of Naess' deep ecology is the *complexity, not complication principle*. It favors the preservation of complex interrelations which exist between parts of the environment, and inevitably, it involves a systems theoretical orientation. Naess' ideal is a complex economy supported by division, but not fragmentation of labor; worker alienation to be overcome by opportunities for involvement in mental and manual, specialized and nonspecialized tasks. There are serious problems of implementation attached to this vaguely sketched scenario, but beyond this, the supporting arguments are also weak, not to say very uncritical in terms of the stated aims of the deep ecology movement. The references to 'soft future research,' 'implementation of policies,' and 'exponential growth of technical skill and intervention,' are highly instrumental statements which collapse back into the shallow ecology paradigm and its human chauvinist ontology. What appears to be happening here is this: the masculine sense of self-worth in our culture has become so entrenched in scientistic habits of thought that it is very hard for men to argue persuasively without recourse to terms like these for validation. Women, on the other hand, socialized as they are for a multiplicity of contingent tasks and practical labor functions in the home and out, do not experience the inhibiting constraints of status validation to the same extent. The traditional feminine role runs counter to the exploitive technical rationality which is currently the requisite masculine norm. In place of the disdain that the feminine role receives from all quarters, 'the separate reality' of this role could well be taken seriously by ecologists and reexamined as a legitimate source of alternative values. As Snyder suggests, men should try out roles which are not highly valued in society; and one might add, particularly this one, for herein lies the basis of a genuinely grounded and nurturant environmentalism. As one eco-feminist has put it:

> If someone has laid the foundations of a house, it would seem sensible to build on those foundations, rather than import a prefabricated structure with no foundations to put beside it. (Pettitt 1982: 20–21)

A final assumption of deep ecology described by Naess is the importance of *local autonomy and decentralization.* He points out that the more dependent a region is on resources from outside its locality, the more vulnerable it is ecologically and socially: for self-sufficiency to work, there must be political decentralization. The drive to ever larger power blocs and hierarchical political structures is an invariant historical feature of patriarchal societies, the expression of an impulse to compete and dominate the Other. But unless men can come to grips honestly with this impulse within themselves, its dynamic will impose itself over and over again on the anatomy of revolution. Women, if left to their own devices, do not like to organize themselves in this way. Rather they choose to work in small, intimate collectivities, where the spontaneous flow of communication 'structures' the situation. There are important political lessons for men to learn from observing and participating in this kind of process. And until this learning takes place, notions like autonomy and decentralization are likely to remain hollow, fetishistic concepts.

Somewhat apologetically, Naess talks about his ecological principles as 'intuitive formulations' needing to be made more 'precise.' They are a 'condensed codification' whose tenets are clearly 'normative'; they are 'ecophilosophical,' containing not only norms but also 'rules,' 'postulates,' 'hypotheses,' and 'policy' formulations. The deep ecology paradigm takes the form of 'subsets' of 'derivable premises,' including at their most general level 'logical and mathematical deductions.' In other words, Naess' overview of ecosophy is a highly academic and positivized one, dressed up in the jargon of current science-dominated standards of acceptability. Given the role of this same cultural scientism in industry and policy formulation, its agency in the very production of the eco-crisis itself, Naess' stance here is not a rationally consistent one. It is a solution trapped in the given paradigm. The very term *norm* implies the positivist split between fact and value, the very term *policy* implies a class separation of rulers and ruled. Devall, likewise, seems to present purely linear solutions—'an objective approach,' 'a new psychology'; the language of cost-benefit analysis, 'optimal human carrying capacity,' and the language of science, 'data on hunter gatherers,' both creep back in. Again, birth 'control programs' are recommended, 'zoning,' and 'programming,' the language of technocratic managerialism. 'Principles' are introduced and the imperative *should* ride roughshod through the text. The call for a new epistemology is somehow dissociated in this writing from the old metaphysical presuppositions which prop up the argument itself.

In arguing for an eco-phenomenology, Devall certainly attempts to bypass this ideological noose—'Let us think like a mountain,' he says—but again, the analysis here rests on what is called 'a gestalt of person-in-nature': a conceptual effort, a grim intellectual determination 'to care'; 'to show reverence' for Earth's household, and 'to let' nature follow 'its separate' evolutionary path. The residue of specular instrumentalism is overpowering; yet the conviction remains that a radical transformation of social organization and values is imminent: a challenge to the fundamental premises of the dominant social paradigm. There is a concerted effort to rethink Western metaphysics, epistemology, and ethics here, but this 'rethink' remains an idealism closed in on itself because it fails to face up to the uncomfortable psychosexual origins of our culture and its crisis. Devall points by turn to White's thesis that the environmental crisis derives from the JudeoChristian tradition, to Weisberg's argument that capitalism is the root cause, and to Mumford's case against scientism and technics. But for the eco-feminist, these apparently disparate strands are merely facets of the same motive to control which runs a continuous thread through the history of patriarchy. So, it has been left to the women of our generation to do the theoretical housework here—to lift the mat and sweep under it exposing the deeply entrenched epistemological complexes which shape not only current attitudes to the natural world, but attitudes to social and sexual relations as well (Salleh 1984a). The accidental convergence of feminism and ecology at this point in time is no accident.

Sadly, from the eco-feminist point of view, deep ecology is simply another self-congratulatory reformist move; the transvaluation of values it claims for itself is quite peripheral. Even the Eastern spiritual traditions, whose authority deep ecology so often has recourse to—since these dissolve the repressive hierarchy of Man/Nature/God—even these philosophies pay no attention to the inherent Man/Woman hierarchy contained within this metaphysic of the Whole. The suppression of the *feminine* is truly an all-pervasive human universal. It is not just a suppression of real, live, empirical women, but equally the suppression of the feminine aspects of men's own constitution which is the issue here. Watts,

Snyder, and Devall, all want education for the spiritual development of 'personhood.' This is the selfestranged male reaching for the original androgynous natural unity within himself.

The deep ecology movement is very much a spiritual search for people in a barren secular age; but how much of this quest for self-realization is driven by ego and will? If, on the one hand, the search seems to be stuck at an abstract cognitive level, on the other, it may be led full circle and sabotaged by the ancient compulsion to fabricate perfectability. Men's ungrounded restless search for the alienated Other part of themselves has led to a society where not life itself, but 'change,' bigger and better, whiter than white, has become the consumptive end. The dynamic to overcome this alienation takes many forms in the post-capitalist culture of narcissism-material and psychological consumption like karma-cola, clown workshops, sensitivity training, bio-energetics, gay lib, and surf-side six. But the deep ecology movement will not truly happen until men are brave enough to rediscover and to love the woman inside themselves. And we women, too, have to be allowed to love what we are, if we are to make a better world.

Bibliography

Devall, Bill. 1980. The Deep Ecology Movement. *Natural Resources Journal* 20 (31): 299–322.

Naess, Arne. 1973. The Shallow and the Deep, Long Range Ecology Movement. *Inquiry* 16: 95–100.

Pettitt, Ann. 1982. Women Only at Greenham. *Undercurrents* 57: 20–21.

Rodman, John. 1977. The Liberation of Nature? *Inquiry* 20: 83–145.

Salleh, Ariel. 1981. Of Portnoy's Complaint and Feminist Problematics. *Australian and New Zealand Journal of Sociology* 17 (1): 4–13.

———. 1983. Ecology and Ideology. *Chain Reaction* 31: 2–21.

———. 1984a. Contribution to the Critique of Political Epistemology. *Thesis Eleven* 8: 23–43.

———. 1984b. The Growth of Eco-feminism. *Chain Reaction* 36: 26–28.

———. 1984c. Whither the Green Machine? *Australian Society* 3: 15–17.

———. 1985. From Feminism to Ecology. *Social Alternatives* 4 (3): 8–12.

Ariel Salleh is Visiting Professor of Culture, Philosophy and Environment, Nelson Mandela University, South Africa; Senior Fellow in Post-Growth Societies, Friedrich Schiller University Jena, Germany; and Research Associate in Political Economy, University of Sydney, Australia. Her transdisciplinary writing is seminal to political ecology as an emerging study of humanity-nature relations. Salleh was Associate Professor in Social Inquiry at the University of Western Sydney for a number of years and has lectured at New York University; Institute for Inter-Cultural Studies, Manila; York University, Canada; and Lund University, Sweden. Her theory of an embodied materialism and meta-industrial labour is developed in *Ecofeminism as Politics* (1997/2017), *Eco-Sufficiency & Global Justice* (ed. 2009), and some 200 chapters and articles in journals such as the *Journal of World Systems Research* (US), *Globalizations* (UK), *Environmental Ethics* (US), *Arena* (AU), *New Left Review* (UK), *Organization & Environment* (US), *Environmental Politics* (UK) and *The Commoner* (UK).

3

Relating to Nature: Deep Ecology or Ecofeminism?

Freya Mathews

Two of our most seminal philosophies of nature, deep ecology and ecofeminism, offer alternative accounts of our relationship with the natural world. Deep ecology tends to take a basically holistic view of nature—its image of the natural world is that of a field-like whole of which we and other 'individuals' are parts. It encourages us to seek our true identity by identifying with wider and wider circles of nature, presenting the natural world as an extension of ourselves, the Self-writ-large. In this view, our interests are convergent with those of nature, and it becomes incumbent on us to respect and serve these common interests.

Ecofeminists, in contrast, tend to portray the natural world as a community of beings, related, in the manner of a family, but nevertheless distinct.[1] We are urged to respect the individuality of these beings, rather than seeking to merge with them, and our mode of relating to them should be via open-minded and attentive encounter, rather than through

Originally published as: Mathews, Freya. 1994. Relating to Nature: Deep Ecology or Ecofeminism? *The Trumpeter: Journal of Ecosophy* 11 (4) (Fall): 159–166.

F. Mathews (✉)
La Trobe University, Melbourne, VIC, Australia

L. Stevens et al. (eds.), *Feminist Ecologies*,
https://doi.org/10.1007/978-3-319-64385-4_3

abstract metaphysical preconceptualization. The understanding born of such encounters should result in an attitude of care or compassion which can provide the ground for an ecological ethic.[2]

Although the tension between these two theories cannot be resolved by merely cutting and pasting them together, I think that a dialectical reconciliation of their respective views of nature can be achieved, though this may result in an irreducibly ambivalent ecological ethic. Such ambivalence may in fact be precisely what an adequate understanding of the ecological structure of reality requires.

In this essay, I begin with an examination of the metaphysical axioms of deep ecology. I argue that these axioms generate a fundamental dilemma for deep ecologists. In attempting to resolve this dilemma, I find I have to give up the ethical conclusions to which deep ecology is normally assumed to lead, and draw instead on an ethical perspective more akin to that found in ecofeminist literature.

The Two Metaphysical Axioms of Deep Ecology

The primary axiom of deep ecology is the thesis of metaphysical interconnectedness. Arne Naess images the natural world as a field of relations. He advocates:

> rejection of the man-in-environment image in favour of the relational total-field image. Organisms as knots in the biospherical net or field of intrinsic relations. An intrinsic relation between two things A and B is such that the relation belongs to the definitions or basic constitutions of A and B, so that without the relation, A and B are no longer the same things. The total field model dissolves not only the man-in-environment concept, but every compact thing-in-milieu concept—except when talking at a superficial or preliminary level of communication. (Naess 1973)

In an early paper, Warwick Fox identifies as the 'central intuition' of deep ecology the idea 'that there is no firm ontological divide in the field of existence … To the extent that we perceive boundaries, we fall short of deep ecological consciousness'.[3] All exponents of deep ecology seem to agree that individuals, to the extent they can be identified at all, are constituted out of their relations with other individuals: they are not discrete substances capable of existing independently of other individuals. The whole is understood to be more than the sum of its parts, and the parts are defined through their relations to one another and to the whole.

The second metaphysical presupposition of deep ecology functions more as a hidden premise—it is not listed as an axiom, as the interconnectedness thesis is, but, so far as I am aware, it is nevertheless taken for granted in all versions of the theory. The presupposition in question is that nature can best look after its own interests, that it is only our interventions in the natural course of events that give rise to terminal ecological disasters. This assumption is implicit in the injunction to let nature take the lead in ecological matters, to minimize our interference in it and to try to shape our own interests to those of nature. It is neatly summed up in Barry Commoner's third law of ecology: nature knows best.

Now let us look at the implications of these two metaphysical assumptions for our relation to the natural world. According to deep ecologists, the fact of our interconnectedness with the rest of nature implies that we are ultimately identifiable with nature; the fact of the indivisibility of reality implicates us in wider and wider circles of being. We should accordingly shed our confining ego identity, and gradually open up to nature at large. The process of achieving the widest possible identification with nature is equated, in deep ecology, with Self-realization: Self-realization is a matter of enlarging one's sphere of identification.

Normative implications are taken to follow hard on the heels of this identification thesis, together with the assumption that nature can and should look after its own interests. For if we are in this sense one with nature, and our interests are convergent with those of nature, then we shall be called upon to defend nature from human interference, just as we are called on to defend ourselves against attack. As activist and deep ecologist John Seed puts it, 'I am protecting the rainforest' develops to 'I am part of the rainforest protecting myself' (Seed 1985). Recognition of our identifiability with nature is taken to entail a commitment to ecological resistance.

The Identification Dilemma

At this point in the argument however, an intractable dilemma raises its head. I shall call it the 'identification dilemma'. If we are identifiable with nature, as the interconnectedness thesis implies, then whatever we do, where this will include our exploitation of the environment, will qualify as natural. Since nature knows best how to look after itself, it follows that whatever qualifies as natural must be ecologically for the best, at least in the long run. In short, if we are truly part of, or one with, nature, and nature knows best, then our depredations of the natural world must be ecologically, and hence morally, unobjectionable.

To this objection, a deep ecologist might reply that although we are ontologically one with nature, we may not consciously recognize this to be the case. In consciousness, we may construct our identity in opposition to nature. Our actions vis-à-vis the environment will then reflect this false consciousness, rather than the underlying ontological fact: we shall be acting as if we were ontologically detached even though this is not in fact the case. Such action may then be regarded as unnatural, in the sense that it does not testify to our actual interconnectedness with the rest of the world.

This reply however would appear to conflate the natural with the true. It may be perfectly natural for consciousness to belie the ontological facts, for there may be adaptive value in its doing so in certain circumstances. After all, there are many species which, though ontologically interconnected with the rest of life (according to the interconnectedness thesis), nevertheless appear to act out of narrow self-interest and exploit the environment to the best of their ability for their own ends. ('Plagues' of locusts and mice spring to mind in this connection; but many species, even in normal circumstances, tread anything but lightly on their lands, relying on the regenerative powers of nature rather than on their own restraint to ensure the continuing health of their environments. The noble elephant is a case in point.) Such a gap between consciousness and the ontological underpinnings of a species' identity may well serve nature's own purposes—it may be part of the long-term ecological scheme of things. If this is the case, then such a gap would be ecologically, and hence ethically, unobjectionable. If we consider it desirable that our consciousness should reflect our true ontological estate, then we cannot claim that this is because such fidelity to ontology is natural; we must rather admit that it is because we value truth. But then there is no reason to suppose that the present self-interested, exploitative behavior of humanity is unnatural; and if it is natural—if it is in accordance with the ways of nature—it cannot, from a deep ecological viewpoint, count as wrong.

In sum, it is plausible to argue, in the light of the interconnectedness thesis, that whatever we do to the environment is natural, and that, since nature knows best, our present despoliation of the environment must in fact be in nature's long-term interests. We might wish to change our ways on our own behalf, recognizing that we are at present orchestrating our own extinction. But we have no grounds for changing our ways on behalf of nature, which is to say, on grounds of ecological morality. To suppose otherwise is in fact to perpetuate the old division between humanity and nature, and with it the old assumption of human supremacism. For to suppose that we can destroy nature is to deny that nature knows best, where this is to admit that we had really better take the rudder after all, and steer nature through this crisis that we have created for it. In other words, to allow that what we are doing to the environment is natural, and yet to insist that it needs to be changed by us, is to deny that nature knows what it is doing; it is subtly to re-usurp control. If we are true to the metaphysical premises of deep ecology, if we accept both our oneness with nature and nature's fitness to conduct its own ecological affairs without our assistance, then we should allow our own evolution to run its 'natural' course, whatever that turns out to be, on the understanding that by doing so we shall be advancing the cause of life on earth. It may well be that our massive impact on the planetary ecosystem is paving the way for an epoch-making transition in evolution—perhaps analogous to the transition from anaerobic to aerobic life in the early stages of the history of life on earth.

The insistence of deep ecologists that we are one with a nature which best knows how to look after itself then does seem directly to imply that we have no ecological nor, hence, moral grounds for intervening in the spontaneous course of human affairs as these affect the environment. This poses a dilemma for deep ecology, since deep ecologists have no desire so to acquiesce in the present regime of environmental degradation and destruction. If they persist—as I have no doubt they will—in exhorting us to engage in active 'ecological resistance', then we have to conclude that there is an inconsistency at the heart of deep ecology.

Holistic and Individualistic Readings of the Two Axioms

If, as environmentalists, we are already committed to ecological resistance, the conclusion of the previous section forces us to re-examine the two metaphysical premises of deep ecology. One or both of them will have to be modified, in some way, if deep ecology is to retain its activist appeal. Let us then review each of these axioms in turn.

The interconnectedness thesis. Is there anything logically amiss with the idea of interconnectedness that is so central to deep ecology, anything that would account for the counterintuitive conclusion to which, when conjoined with the thesis that nature knows best, it was found to lead? I think the problem with this thesis, in the present connection, is not that its interpretation within deep ecology is in any way logically flawed, but merely that it is partial.

Deep ecologists have, in the main, given the idea of interconnectedness a holistic reading; they have taken it to mean that nature, as a metaphysical whole, is logically prior to its parts, and that the identity of each part is functionally determined by way of its relation to the whole. They concede a degree of autonomy to individuals, but ultimately they view that autonomy as apparent only, without fundamental ontological significance. Different exponents of deep ecology offer slightly different accounts of the ontological status of individuals (and hence of the relationship between self and nature).[4] However, despite these differences, the holistic emphasis remains marked: the viewpoint of the individual must, in one way or other, be given up in favor of the viewpoint of the whole. We and all other individuals are ultimately seen as in some sense 'one with' nature.

It is arguable however that this reading of the interconnectedness thesis captures only one side of its meaning. If a systems-theoretic approach is adopted, it is possible to see interconnectedness as entailing the identities of both wholes and individuals. From a systems-theoretic viewpoint, the world (particularly the biological world) appears as a field of relations, a web of interconnections, which does indeed cohere as a whole, but within which a genuine form of individuation is nevertheless possible. An individual is, from this viewpoint, an energy configuration or system which maintains itself by way of its continuous interactions with its environment. Since it is only able to maintain its integrity by way of this continuous give and take with the environment, its existence is a function of its relations, its interconnections. But since these interactions do indeed enable it actively to maintain its integrity, it does enjoy a genuine, though relative, individuality. In this way, the world may be seen as both a seamless whole and a manifold of individuals.[5]

On this reading then, metaphysical interconnectedness implies an irreducible ontological ambivalence at the level of individuals: individuals are, in this scheme of things, analogous to the 'wavicles' of quantum mechanics. In quantum mechanics, light is analyzed in terms of these wavicles: looked at from one point of view, a ray of light manifests as a stream of particles (photons), while from another point of view, it manifests as a wave phenomenon (a pattern in a field). Light cannot be reduced to either photons or field. Ontological ambivalence is thus intrinsic to its nature.

Under the sway of the interconnectedness thesis, deep ecology tends to view the natural world from the holistic perspective exclusively, and therefore considers individuals as field-like rather than as particulate. This one-sided reading of the interconnectedness thesis inevitably also affects its reading of the principle that nature knows best. The principle that nature knows best will be understood to mean that nature knows best for itself as a whole, but it is not taken to imply that nature knows best for the individuals that are its elements. Reading the principle in this latter sense raises obvious questions about its validity. Let us look at the principle in the light of this double reading, and consider whether it can be retained.

The thesis that nature knows best. The principle that nature knows best implies that nature is the best servant of its own interests, and therefore that, from the viewpoint of environmental ethics, whatever nature does is right. It follows from this that the natural order is a moral order, that within this natural order everything ultimately turns out for the best, so far as nature is concerned. Can this assumption be defended? In order to answer this, we need, as I have pointed out, to look at the principle under both its holistic and its individualistic interpretations. I shall argue that under the holistic interpretation, the natural order is indeed a moral order, but that under the individualistic interpretation, it is not.

The answer to the question whether nature knows best, when nature is viewed under its holistic aspect, depends to some extent on the empirical question of whether or not we, or any other particular life form, have the capacity to extinguish life altogether on the planet. On current evidence this appears to be unlikely: it is widely believed that even full-scale nuclear holocaust would fail to eliminate microbial life forms and that the adaptations of these life forms to the new conditions would usher in a new evolutionary epoch. In light of this assumption that the demise of one order of life creates an opportunity for another, I think we can say that, from the viewpoint of the whole, nature inevitably works toward its own good.

Nature—understood under its holistic aspect—knows best not only in the sense that it is capable of looking after its own interests; it appears to know best in a wider moral sense as well, since the ecological order not only secures its own self-perpetuation, but also appears to exemplify both justice and generosity. Such ecological justice consists, in the first place, in the fact that ecological 'transgressors' pay for their ecological 'transgressions' by being selected out of existence; and it consists, in the second place, in the fact that such self-elimination of actual individuals provides possible individuals with their opportunity to gain entry into the actual world. Such perfect impartiality between the actual and the possible must surely represent the acme of justice! If it is objected that it is scarcely just to condemn an entire ecosystem to extinction on account of the ecological 'transgressions' of one of its elements, it must be remembered that from the holistic point of view there is no absolute distinction between an element and its ecosystem. The various elements of an ecosystem are merely different expressions of its own intrinsic logic or theme. It makes no sense, from this holistic perspective, to say that we, as ecological deviants, are endangering our otherwise ecologically viable ecosystems, or the ecologically innocent elements of those ecosystems. For if we are deviant, so are the ecosystems with which we are holistically or internally related, and so too are all the elements of those ecosystems. If we deserve to be selected out for our mistakes, so too does the ecosystem, or even the entire order of life, which defines us.

From the holistic point of view then, the natural order is arguably an order of justice and as such qualifies as a moral order in a richer sense than that implied in the original maxim that nature knows best. Lest such a moral order seem too stern for us to countenance, however, there is, as I remarked earlier, a second way in which the natural—still viewed from a holistic perspective—is equivalent to the right. The moral significance of nature, understood in this second sense, resides in its boundless generosity. Etymologically, 'nature', as Holmes Rolston III points out, is derived from the Latin *natus*, meaning birth. Nature is the source, the wellspring, of life, and life is, after all, an entirely gratuitous gift, owed to no one. 'When nature slays', says Rolston, 'she takes only the life she gave … and she gathers even that life back to herself by reproduction and re-enfolding organic resources and genetic materials and produces new life out of it' (Rolston 1979). Because nature does not favor those who have life over those who do not, life is dealt out lavishly: the dispensability of the actual is a necessary condition for this lavishness. Nature is not only just, but infinitely generous. The natural order then, viewed from the holistic perspective, is moral not only in that it secures the long-term good of nature, but also in its justice and its generosity.

When nature is examined from the individualistic rather than the holistic viewpoint however, does it still qualify as a moral order? Is the natural still the right? We have seen that, from the point of view of the whole, individuals are generously given life and justly sacrificed that the gift of life might be passed on. As long as we are (quite properly) identifying with the whole, we can appreciate both the effectiveness and the justice of this arrangement, and concur in the price that is paid for it. When we (equally properly) identify ourselves as individuals however, we are likely to see things differently. Nature no longer appears to know best, if by its 'knowing best' we mean that it is capable of looking after the interests of individuals. Nor does it appear as just: the situation of actual individuals is importantly different from that of possible individuals. As actual individuals we have actual interests, urgent needs and desires; we can suffer, and suffer terribly. There is neither justice nor generosity in trading in actual individuals for possible ones, from this perspective. The stern, though admittedly, life-giving 'plan' of nature-as-a-whole then has less to commend it from down here. Nor is it only *our* fate which assumes a larger moral significance from this perspective: that of other actual individuals does likewise. Fellow-feeling for them, familiarity with the imperative which drives them, identification with the shivering vulnerability that their actuality implies, gives rise to concern, to a moral interest in their plight.

Ironically then the impulse to resist the progressive destruction of the present order of life springs not, as deep ecology claims, from our identification with nature-as-a-whole—though that identification is perfectly proper, in light of the holistic interpretation of interconnectedness—but rather from our commitment to our individuality. It is as individuals that we feel concern for other individuals. In defending non-human beings against human depredations, we may even in a sense be resisting the greater moral order, the grand order of ecological justice. The compassion which forms the basis of our environmental ethic, from this individualistic point of view, is a function of our finitude rather than of our cosmic self-realization. In securing the conditions for the ongoing unfolding of life, nature (in its holistic aspect) is morally more far-sighted than we; in the name of compassion we seek to block that unfolding by clinging to those individuals which already exist, out of a sense of solidarity with them. As individuals we give our allegiance to individuals, if necessary even against the moral requirements of nature-as-a-whole.

Deep Ecology and Ecofeminism: Complementary Perspectives?

This view of the basis of environmental ethics is much closer to ecofeminism than to deep ecology. Ecofeminism is by no means a position or a theory, but simply a fairly open field of inquiry, but it could nevertheless be taken to subscribe to the interconnectedness thesis.[6] It tends to interpret interconnection in the individualistic rather than in the holistic sense: nature, from the ecofeminist perspective, is a community of beings, related, in the manner of a family, but nevertheless distinct. We are urged to respect the otherness, the distinct individuality of these beings, rather than seeking to merge with them, in pursuit of an undifferentiated oneness.

Since ecofeminism does not identify us directly with nature-as-a-whole, it does not fall foul of the identification dilemma. In other words, since it does not define us as identifiable with a monolithic nature, it does not have to see our destruction of the environment as a case of nature 'destroying' itself, where seeing our action in this way renders it morally unobjectionable. On the contrary, since it sees us as related to nature as to the members of a community or family, to whom the proper attitude is one of familial consideration and care, born of an empathetic understanding made possible by our common origins, or our mutually defining relations, ecofeminism is able to condemn our abuse of the environment outright: this is no way to treat one's family! So for ecofeminism, concern for nature is the product of a re-awakening to our kinship with our individual non-human relatives; it is grounded in our individuality, rather than in any kind of cosmic identification, and it springs out of a sense of solidarity with our fellow beings.

It seems to me, as I indicated at the outset, that ecofeminism and deep ecology, with their complementary interpretations of the interconnectedness thesis, each captures an important aspect of our metaphysical and ethical relationship with nature. For if reality is indeed internally interconnected, if it does consist in a web of relations, then, as I explained earlier, it may be seen as both a whole and a manifold of individuals. From the viewpoint of the whole, it does appear to qualify as a moral order, though from the viewpoint of the individual, it does not. Since I claim both these viewpoints need to be taken into account in our attempt to determine how we should relate to nature, we find ourselves committed in the end to an irreducible moral ambivalence consisting of compassionate intervention on behalf of nature on the one hand, and enlightened acquiescence in the natural tide of destruction on the other. In accepting this ambivalence, we discover on the one hand that it is our humanity—our very finitude and limitation—rather than any grand plan in the stars that impels us to act on behalf of our embattled fellow creatures. In this way the moral loftiness of deep ecology is brought down to the ground, rendered human. But on the other hand we discover that our compassion—the value taken for granted by ecofeminism—is not beyond moral question either. In light of the grand plan that *is* in the stars, compassion is seen to come down to our love of the familiar, our solidarity with the things that remind us of ourselves.

The recognition that our grounds for ecological resistance lie in our humanity, rather than in our Self-writ-large, or in the stars, is particularly important for environmentalists, I think. For many environmentalists, face to face with the heartbreaking consequences of human rapaciousness become embittered toward humankind and come to see our species as a curse upon the earth. Out of such a relapse into dualistic thinking, no true healing or affirmation of life can come. To recognize that our humanity is the wellspring not only of a consuming destructiveness but also of the precious compassion which counters it may be a redeeming thought, which will help to lead us out of the moral impasse created by the divorce between humanity and nature. It is to the roots of this divorce in dualistic patterns of thought that I shall now turn.

Dualism: Deep Ecological and Ecofeminist Responses

In this final section I would like to explore the ways in which deep ecology and ecofeminism, despite their contrasting (though on my account complementary) ethical perspectives, are inexorably at many points drawn into each other's orbit by the force of their common effort to escape the dualism that grips our Western conceptual framework.

Deep ecologists, as we have seen, assert that we as human beings are identifiable with nature-as-a-whole, but according to my argument, they then generate an inconsistency by insisting that, once we have recognized this identifiability, we should ally ourselves with nature against humankind. In other words, they re-assert a sharp division between humankind and nature. If deep ecology is to be consistent, I have argued, it should give up this division and the struggle to which it gives rise, and surrender to the spontaneous course of human affairs. Since I do not think this is a conclusion which most deep ecologists would be prepared to accept, I shall not refer to this position of resignation simply as 'deep ecology', even though it is, according to my argument, truer to the premises of deep ecology than is the view which normally goes by that name. I shall instead refer to this position as 'cosmic ecology', or perhaps simply 'the cosmic view'. According to cosmic ecology then, our identification with nature-as-a-whole entails a moral acquiescence in all human action, insofar as it impinges on the environment, since our actions are now seen as manifestations of a cosmic order which is, so far as the environment is concerned, inherently moral.

From the viewpoint of ecofeminism, we as human beings are not identifiable with nature understood in a monolithic sense; rather we are members of the wider family of life. In recognition of the ties of kinship between ourselves and the other members of this family, we are motivated to treat those others with care and consideration. This may on occasion involve protecting non-human members from their human relatives, but the struggle that ensues will not be of the us-against-them variety, but will rather be many-sided. It will involve resisting the actions of some members in some circumstances, while being prepared to affirm the actions of those same members in others. Such a struggle will resemble the struggle that a mother may face within her family—restraining outbreaks of aggression among her offspring, while not allying herself with one family member against another. We who feel loyalty both to our human and to our non-human relatives are in much the same position as this mother; our task is to restore the set of relationships which will enable the family to function as a healthy system.

Cosmic ecology then appears to prescribe quietistic surrender to whatever is the case, while ecofeminism advocates many-sided negotiation for the sake of accommodating all our relations. Despite this contrast in their prescriptive outcomes however, the two views, as I indicated at the beginning of this section, converge in certain vital respects. To see this, let us begin by looking more closely at the implications of the cosmic view.

Can we really accept the idea, implicit in the cosmic view, that human life, however lethal in its intent and its impact on the natural world, is nevertheless tributary to the ultimate moral order? It goes painfully against our grain, as environmentalists, to concede that the bulldozer and its driver are contributing to the moral order just as effectively as the forest is. Nevertheless, it is, I believe, important for environmentalists to concede this, since the typical deep ecological reverence for untouched nature—idealized in the concept of wilderness—is rooted in the very same dualistic understanding of the world that, by setting humankind above and beyond nature, paved the way for the ecological crisis. If we make a fetish of untouched nature, then we are implicitly reinforcing this dualistic view. To maintain this division—albeit reversing the values that dualistic thinking has traditionally assigned to nature and to humankind respectively—is, as I have explained at length, to contradict the basic metaphysical premise of deep ecology, namely, the interconnectedness thesis.

In conceding that nature is reflected in the bulldozer and its driver just as faithfully as it is in the forest, we are in fact transforming the traditional environmentalist image of nature. For many environmentalists, as I have remarked, true nature manifests itself in inverse proportion to its proximity to human activities or interventions. In other words, nature is in its truest state in wildernesses or remote regions. We can accordingly expect to experience the loss of nature most acutely in those places where humanity is most concentrated, as in the cities, the great metropolises of the late twentieth century. This assumption of course cannot be sustained in the light of the cosmic view, with its characterization of the human order as an instance of the natural order. The city itself, from this point of view, becomes a teeming locus of nature, a field of relations inevitably organizing itself into increasingly diverse and complex forms, where this efflorescence of new forms takes place not at a biological but at a cultural level.

Recognition of this suggests the further jolting insight that nature may not after all be confined to biology—that while it may have invented species as a vehicle for diversity and complexity, other forms of diversity and complexity might express its underlying essence or telos just as well. It is we, rather than nature, who are fixated on species, just as it is we, rather than nature, who agonize over the fate of individuals. Maybe nature can realize itself through emergent levels of culture, perhaps even—who knows?—through emergent levels of computer functioning. Given time, nature will invariably create the order, the endlessly elaborated and modulated themes, that are so beautifully but perhaps contingently expressed in the biological and ecological life of this planet.

Looking at the city from the cosmic point of view then, we might register an intensification of the pulse of life there. Perhaps here, in the heart of the metropolis, nature is at its wildest. Certainly life is fast and full and dangerous in these streets, taut with uncertainty and unexpectedness. Perhaps as the wilderness retreats across the continents, its spirit returns, bright and sexy and violent, into our very midst. From this point of view, nature cannot die at our hands—everything we do merely constitutes its further unfolding. From the recognition that we and all our activities and contrivances are an expression of nature then, a new image of nature does indeed emerge. We can expect to discover its underlying Tao in the love-and-struggle-and-crimefilled streets of London or Tokyo just as surely as on the Siberian taiga or in the deserts of western Australia.

The same argument can be applied in relation to our artifacts, our technologies. The instruments of ecological destruction—the bulldozers, oil drills, missiles, H-bombs—are generally abhorred, even demonized, by environmentalists. To adopt the cosmic view however, and to recognize our true identity with nature, is to recognize that these technologies are all instruments of the natural order, on a par with tusks and venom, cyclones, landslides and ice ages. They are fashioned out of terrestrial materials by one of the earth's species and set in motion by that species' telos. If we truly honor the earth, we should honor these forms that have always been latent within it, and we should honor these emerging potentialities of its nature. Besides, since it is our technology which mediates our relationship with the world, we cannot honor the world if we despise our technology. In spiritual terms, we need, like the primal peoples so admired by deep ecologists, to locate the sacred not merely in the cosmos, but in the technology which discloses the cosmos to us.[7] Many of those primal peoples attributed an indwelling spirit to their artifacts. The latter were enchanted, charged with a life and destiny of their own, just as the wider world was. From the cosmic point of view, we need urgently to sacralize our own dangerously secular technologies, if we are to respect the world that these technologies open up to us.

To be prepared to accept as natural and hence to respect—perhaps to sacralize—our cities and our technologies of destruction is to respect and re-enchant the nature that we actually inhabit—as opposed to the nature that exists in some remote region which we may never visit, some world locked away in a reserve or fenced against human intrusions. It is within our own everyday world that we must forge our relationship with nature, and perhaps rediscover the sacred.

As it happens, these implications of cosmic ecology echo certain of the sentiments that ecofeminists have recently been expressing. Irene Javors, for instance, has said, in the idiom of feminist spirituality,

> The Goddess lives in the city. She is present in all her manifestations. However, we have great difficulty dealing with her as Hecate/Kali, the destroyer/crone. We fear the 'gifts' that she brings us—age, change, deterioration, decay, death. She is an alchemist who finds the seeds for new life within the compost heap of decomposing forms. We fear her and run from her dark side; by so doing, we blind ourselves to her holiness. (Javors 1990)

And another ecofeminist writer has recommended the resacralization of our technologies in the following terms:

> I believe it is time to create new songs of acknowledgement as well as ceremonies that include metals, petrochemicals and fossil fuels, electricity, modern solar power systems, and water power systems. I also believe it is very important to make sacred, to acknowledge the new ways and elements in our lives—from nuclear power (which is buried in our earth and activates our Sun) to plastics to computers. It is time now, again, for the entire world to honor these Spirits, these new molecular forms, to restore harmony and balance to our out-of-control systems and in particular, to our modern technologies. (Sanchez 1989)

Why is it that ecofeminists are beginning to enter the same spiritual terrain as the cosmic version of deep ecology? The argument behind these ecofeminist sentiments is quite different from the argument that leads to the cosmic view, but the two arguments are to some extent convergent. The argument which led to the cosmic view was, as we have seen, that overcoming the dualistic division of humankind and nature entailed accepting human destructiveness as natural and therefore as morally unobjectionable. The ecofeminist argument centers on dualism too, but ecofeminists offer a much more systematic analysis of dualistic patterns of thought than deep ecologists do. From the ecofeminist point of view, dualism constitutes a full-blown ideology which interprets the world in terms of dichotomous pairs of qualities, such as active/passive, light/dark, mind/body, reason/emotion and culture/nature. Not only are the qualities that appear in these pairs of opposites dichotomized, in this dualistic scheme of things, they are also hierarchically ordered: within each of the above pairs of opposites, the left-hand term is invariably regarded as 'higher' than the term on the right. The reason for this, according to the ecofeminist analysis, is that the terms on the right are defined via their association with the feminine, while those on the left are identified with the masculine. The entire system exists for the purpose or legitimating the inferiorization of the feminine and all things traditionally associated with it.

From the ecofeminist perspective then, the split between humanity and nature that deep ecology seeks to heal is only one instance of a system of dualistic constructions that are psychosexual in origin and political in purpose. Hostility to nature is built into the very foundations of this patriarchal ideology, and the entire ideology must be dismantled if humanity and nature are to be re-integrated. In other words, we cannot set about uniting humanity with nature without at the same time effecting the demolition of this entire system of dichotomizations, including the original dualistic construction of masculine and feminine.

The ecofeminist critique of dualism then has been more concerned with rehabilitating—re-honoring—all the repressed terms in this entire system of pairs of opposites than with simply demonstrating the inextricability of humankind from nature. Within the dualist framework, it has of course been primarily the body, the emotions, eros, nature and the feminine that have been repressed. For this reason, ecofeminists have typically been concerned to celebrate these 'earthy' things. But death, decay and destruction are further aspects of 'earthiness' and have accordingly also been repressed. Ecofeminists are on the verge of pointing out that most environmentalists perpetuate this form of repression in their refusal to accept either the destruction of the non-human world or the human instruments and centers of this destruction, where this still really amounts to a refusal to accept the dark side of nature itself. I am not sure that any ecofeminist has actually said that overcoming dualism involves embracing the destruction of the natural world, but this may in fact be a logical conclusion of the ecofeminist critique of dualism. By way of this rather different route then, ecofeminism appears to converge with the cosmic view in its conclusion that the destruction of the natural world at human hands cannot be regarded as an absolute evil.

In these different ways, ecofeminism and the cosmic version of deep ecology appear to be pointing to what might be an important truth for environmentalists, namely that we cannot save the world without first acquiescing in its loss. The belief that we can save the world rests on the very same assumptions that underlie our attempts to destroy it, these being the assumptions that, in the first place, we are in some sense bigger than the system (and are therefore capable of both destroying and saving it), and that, in the second place, death, destruction and extinction are in any case wrong, and not to be tolerated. Only when we accept the dark side of nature, and see it exemplified in our own destructiveness, can we truly begin to honor nature. And only when we honor it, understanding its dark side, will we be capable of approaching the world in a spirit of receptive encounter, for it is presumably, as many feminists have argued, our fear of this dark side, particularly the prospect of our own mortality, which underlies our drive to conquer, control, dominate and even destroy the world. Ironically then, it is by accepting and honoring the forces of destruction that we are freed from the impulse to destroy.

If strands not only of deep ecology but also of ecofeminism lead to an acquiescence in human destructiveness, an acquiescence that is ultimately the key to transcending that destructiveness in ourselves, does it follow that no grounds remain for ecological resistance, for the protection of non-human life from human exploitation? I think not. The ecofeminist rehabilitation of the dark side of nature has to be set in the context of its ethic of care and kinship. We may accept the dark side, the inevitability, even sacredness, of death and destruction, and yet continue to look out for our kin, continue to protect those for whom we care, in the way that I explained at the end of the previous section. To stand vigilant guard over those whom we love is not necessarily to try to cheat death, nor does it necessarily involve the repression of 'the dark face of the goddess'. A balance must be found between the cherishing of life and the honoring of death. To cherish life need not entail subduing and taking control of nature, and to honor death need not entail abandoning ourselves and all our loved ones to the winds of chance. Our task is to maintain—and perpetually to renegotiate—the dynamic ambivalence which is the life-blood of a healthy morality, a living spirituality. Our acquiescence in mortality may thus lead us to a deep attunement to the terms of life, without in the process committing us to quietism. We need to only concede that our interventions on behalf of our fellow beings spring not from enlightenment but from a homely and humble and all-too-human love of kin. 'Enlightenment' consists in the ability to tolerate without bitterness and despair the failure of these interventions, should they indeed fail; for it is only when we are truly capable of this that we will have rooted out our own impulse to conquer and control the world, our impulse to reshape the world closer to the heart's desire.

Notes

1. Jim Cheney brought this point out very clearly in his 1987 article 'Ecofeminism and Deep Ecology'. *Environmental Ethics* 9(2). It is also explored extensively in Val Plumwood. Spring 1991. Nature, Self and Gender. *Hypatia* 6(1). However, as ecofeminism is not typically expounded systematically as a philosophy, other views of nature are also represented in ecofeminist works. Conversely, the view of nature that I have here identified as ecofeminist is also espoused by writers who make no reference to feminist theory at all. See for instance J. Baird Callicott's account of American Indian views of nature in 'Traditional American Indian and Western European Attitudes Toward nature: an Overview'. 1989. *In Defense of the Land Ethic*. Albany: SUNY Press. See also Callicott's book on multicultural environmental ethics. 1994. *Earth's Insights*. Berkeley: University of California Press. Both Callicott and Aldo Leopold, the architect of the land ethic Callicott is concerned to defend, tend to view nature as a community of natural elements and beings, but both also seem to adopt a holistic interpretation of community for ethical purposes, where this would run counter to the ecofeminist tendency. I am not really

concerned to discuss deep ecology and ecofeminism per se here, but rather a certain complex of issues which are central but not exclusive to these two positions. The issues in question concern the relative merits of the individualistic and holistic views of our relationship to nature. An author who has recently addressed these issues without reference to either deep ecology or ecofeminism is Robert W. Gardiner. 1990. 'Between Two Worlds: Humans in nature and Culture'. *Environmental Ethics* 12 (4).

2. Evelyn Fox Keller develops a sophisticated argument along these lines in 1985. *Reflections on Gender and Science*. New Haven: Yale University Press.
3. In his later work, Fox has made more room for a relative form of individuality in his ecological metaphysic. See 1990. *Towards a Transpersonal Ecology*. Boston: Shambhala.
4. Val Plumwood identifies three versions of the deep ecological account of the relationship of self to nature. She calls them the 'indistinguishability account', the 'expanded self' account and the 'transcended or transpersonal self' account. Although there are indeed certain distinctions to be made among these three positions, it seems to me that they all involve basically holistic interpretations of interconnectedness, since they all point to the substitution of a greater Self for the normal self understood as ego or individual. See Val Plumwood. 1993. *Feminism and the Mastery of nature*. New York: Routledge.
5. This argument that the relational nature of systems entails both individuality and holism is developed in my book: 1991. *The Ecological Self*. London: Routledge.
6. This is evident in the web imagery which is so central to ecofeminism, and which appears in a number of ecofeminist titles, for example, J. Plaskow and C. Christ, eds. 1989. *Weaving the Visions*. New York: Harper and Row, and I. Diamond and G. F. Orenstein. 1990. *Reweaving the World: the Emergence of Ecofeminism*. San Francisco: Sierra Club Books. In the latter work, the editors, in their Introduction, characterize the early ecofeminists as those feminists who 'affirmed and celebrated the embeddedness of all the earth's peoples in the multiple webs and cycles of life'.
7. The comparatively easygoing attitude of certain native peoples in this respect, unfettered as they are by hard-and-fast (dualistic) distinctions between what qualifies as natural (and hence sacred) and what does not, is illustrated by a point made by my colleague at La Trobe, Raj Bessarib, concerning a 'dreamer' of the Sardi people in the Kimberley region of Western Australia. This story-teller of the dreamtime, Billy Ahchoo, includes a 'dance of the motorboat' in his repertoire of dreaming dances.

Bibliography

Fox, Warwick. 1984. Deep Ecology: A New Philosophy of Our Time? *The Ecologist* 14: 194–200.

Javors, Irene. 1990. Goddess in the Metropolis: Reflections on the Sacred in an Urban Setting. In *Reweaving the World: The Emergence of Ecofeminism*, ed. I. Diamond and G.F. Orenstein. San Francisco: Sierra Club Books.

Naess, Arne. 1973. The Shallow and the Deep, Long-Range Ecology Movement. *Inquiry* 16: 95–100.

Rolston, Holmes, III. 1979. Can We and Ought We to Follow nature? *Environmental Ethics* 1: 7–30.

Sanchez, Carol Lee. 1989. New World Tribal Communities. In *Weaving the Visions*, ed. J. Plaskow, C.J. Plaskow, and C. Christ. New York: Harper and Row.

Seed, John. 1985. Anthropocentrism. In *Deep Ecology: Living as if Nature Mattered*, ed. B. Devall and G. Sessions. Salt Lake City: Peregrine Smith.

Freya Mathews is Adjunct Professor of Environmental Philosophy at La Trobe University. Her books include *The Ecological Self* (1991), *Ecology and Democracy* (ed., 1996), *For Love of Matter: A Contemporary Panpsychism* (2003), *Journey to the Source of the Merri* (2003), *Reinhabiting Reality: Towards a Recovery of Culture* (2005), *Ardea: A Philosophical Novella* (2016) and *Without Animals Life Is Not Worth Living* (2016). She is the author of over 70 articles in the area of ecological philosophy. Her current special interests are in ecological civilization; indigenous (Australian and Chinese) perspectives on 'sustainability' and how these perspectives may be adapted to the context of contemporary global society; panpsychism and the critique of the metaphysics of modernity; and wildlife ethics in the context of the Anthropocene. In addition to her research activities she manages a private biodiversity reserve in northern Victoria. She is a fellow of the Australian Academy of the Humanities.

4

Women and Nature Revisited: Ecofeminist Reconfigurations of an Old Association

Kate Rigby

The term *écofeminisme* is said to have been first coined in 1974 by radical French feminist Françoise d'Eaubonne. Identifying the underlying cause for the twin crises of overpopulation and overproduction—somewhat reductively—in the age-old patriarchal domination of women, d'Eaubonne called upon feminists to wed their cause to that of the environment and lead the way into a postpatriarchal, genuinely 'humanist', and ecologically sustainable future (d'Eaubonne 1974: 213–252).[1] Over the past 24 years following the publication of *Le Feminisme ou Le Mort* the connections between the position of women and the fate of the earth have been explored in a number of theoretical directions and arenas of action. As the three books under discussion here amply demonstrate (Merchant 1996; Mellor 1997; Salleh 1997), ecofeminism has truly come of age, both as a theoretically sophisticated form of critique and as a global movement of resistance

Originally published as: Rigby, Kate. 1998. Women and Nature Revisited: Ecofeminist Reconfigurations of an Old Association, *Arena Journal* 12: 143–169.

K. Rigby (✉)
Bath Spa University, Bath, UK

Literary Studies, Monash University, Melbourne, VIC, Australia

L. Stevens et al. (eds.), *Feminist Ecologies*,
https://doi.org/10.1007/978-3-319-64385-4_4

and renovation, linking struggles against environmental degradation with the endeavour to overcome social domination, above all on the basis of sex/gender, but also increasingly in terms of 'race' and class.

Carolyn Merchant, Mary Mellor, and Ariel Salleh are among the most prominent socialist ecofeminists in the English-speaking world, from North America, Britain, and Australia, respectively. While Merchant's book is largely comprised of revised versions of earlier work from the 1980s and early 1990s, the one exception being the chapter on ecofeminism in Australia, it provides a useful introduction both to Merchant's own valuable historical examinations of particular metaphorical and material connections between women and nature in the past, and to the various streams within contemporary ecological feminism. As Freya Mathews has observed, 'ecofeminism is by no means a position or a theory, but simply a wide open field of enquiry' (Mathews 1994: 62). Mellor in particular examines crucial points of difference, and agreement, between ecofeminists, as well as between ecofeminism and other feminisms and ecologisms, with considerable theoretical rigour. Both her book and Salleh's do much to advance ecofeminist thinking, especially in addressing the significance of women's position in relation to the conditions of human embodiedness and ecological embeddedness, and in relation to earlier, especially Marxist, theories of social domination and transformation.

Generally speaking, what distinguishes ecofeminist thinking from an environmentally concerned feminism, or, conversely, a pro-feminist environmentalism, is the claim that there is some kind of inherent or structural connection between the patriarchal domination of women (and, in the view of some theorists, other socially oppressed groups) and the ecologically destructive exploitation of the earth. However they may differ in grounding that connection, in asserting its significance all ecofeminists necessarily reaffirm, to some extent, that link between women and nature which liberal and socialist feminists from Simone de Beauvoir onwards have been at pains to sever. According to de Beauvoir's powerful critique in *The Second Sex* (1968), the association of Woman and Nature was a key element in patriarchal ideology which served to legitimize women's exclusion from the public sphere and their confinement to the realm of 'mere' reproduction, centred on the home. Full equality, for de Beauvoir, was premised upon women's liberation from this emersion in the reproductive sphere through the transcendence of their own sexed bodies, for it was above all women's capacity to give birth which underlay both the feminine imaging of Nature and the biologistic reduction of Woman to womb. While de Beauvoir's analysis of the ambivalences underlying men's view of both women and nature was insightful, as Mellor observes, she remained ultimately uncritical of male-dominated culture and dismissive of 'the problems of embodiment' (Mellor 1997: 79). Shulamith Firestone, writing 20 years later than de Beauvoir, was equally insistent that the emancipation of women necessitated their liberation from the womb—a project which in her view was now potentially realizable through the technologization of reproduction and the socialization of child-rearing in the context of a thoroughgoing socialist transformation of human relations (Firestone 1970). For women (and indeed men) whose feminism had been forged on the anvil of this rationalist egalitarian tradition—whether or not they subscribed fully to Firestone's extreme techno-socialist optimism, and even if they had in the meantime developed a less hostile attitude to motherhood—the ecofeminist revaluation of the association between women and nature looked like a politically dangerous fall back into the mirror of patriarchal projection, which could only play into the hands of those reactionary forces who would like to see women once more in their 'proper place'.

While the risk of a reactionary reinscription of the woman-nature connection is forever hovering menacingly over the ecofeminist project, I would agree with ecofeminists such as Merchant, Mellor, and Salleh that this should not necessitate a retreat into what Val Plumwood terms the feminism of 'uncritical equality' (1993). In Plumwood's analysis, egalitarian feminism of this kind is blind to the ways in which the patriarchal 'master model' of the human has been formed in the context of gender, class, race, and species domination: to be 'human', on this model, is to be defined in opposition to 'nature' and its cognate terms (the feminine, the subaltern, the primitive, the body, etc.) (1993: 22). It is this model of the human that has underpinned the flawed, ultimately selfdefeating project of the domination of nature which has assumed a particularly virulent form in western industrial modernity and which now threatens the ecological basis of all human life (not to mention that of millions of other species) on this planet. For women, and other Others, to now uncritically assimilate themselves to this 'master model' might mean to forego the opportunity to develop new forms of knowledge and selfunderstanding, incorporating less exploitative patterns of human to human and human to nature relationships. It is certainly true that the affirmation of a 'community of fate' between women and nature has proven dangerous in the past, most notably in the eugenically oriented Nazi cult of Aryan motherhood. Today, however, as I have argued elsewhere, an even graver danger is presented by the failure to acknowledge the community of fate that exists between humanity and the earth (Beinssen-Hesse and Rigby 1996: 100). In seeking to redress this failure, ecofeminists have argued, the experiences and perspectives of women, among others who have been similarly marginalized in relation to the master model of the human, and who tend to suffer disproportionately from the results of environmental degradation, might prove particularly valuable.

It is indicative of the extent to which ecofeminist theory has remained vitally intertwined with women's grassroots political engagement that Merchant, Mellor, and Salleh all situate their own analyses in the context of both the history of women's environmental activism and the subsequent development and diversification of ecological feminist thought. Indeed, more than a third of Merchant's book is explicitly concerned with 'practice', not only in the final section of that name, which comprises area studies of women and the environment in the United States, Sweden, and Australia, but also in the preceding chapter on women and the US progressive conservation movement in the early years of the twentieth century. Here, in an article that was originally published in 1984, Merchant seeks to redress the gender bias of earlier histories of this movement in telling of the energy and dedication of some of the thousands of middle-class women whose tireless efforts in a number of organizations were highly successful in bringing many areas of perceived ecological and aesthetic value under protection, and in initiating legislation to halt pollution, protect watersheds, and preserve endangered species (Merchant 1985). Merchant's discussion of the women's conservation movement is nonetheless critically reflective, not blindly adulatory. In particular, she draws attention to the fact that these women's environmental work was made possible by the leisure afforded to them by the capitalist gender division of labour, which they themselves were ideologically committed to uphold, justifying their activities in terms of the conservation of 'true womanhood', the home, and future generations of (white, middle-class) American children. It was, however, in large part on account of this division of labour that women were effectively excluded from the American conservation movement when it became professionalized on the eve of the First World War. Moreover, as Merchant observes, although these early women conservationists were feminist and progressive 'as activists in the public interest', and in their opposition to the dominance of commercial concerns, they were 'predominantly conservative in their desire to uphold traditional values and middle-class lifestyles rooted in these same material interests' (Merchant 1996: 136).

By comparison with the women's progressive conservation movement, Second Wave women's environmental activism—although perhaps not without its own contradictions—has become far more broadly based geographically, as well as in terms of class, 'race', and indeed sexual orientation,[2] and it is potentially far more radical in its challenge to prevailing social relations and commercial interests. In most western countries, women activists have mobilized above all around the issues of nuclear energy and weaponry and chemical toxins, although as Salleh's rather breathless run-down of 'ecofeminist actions' around the world indicates, there seems to be no environmental issue with which some women somewhere have not been concerned.[3] As far back as 1962, as Salleh notes, 'an astonishing series of law suits against the corporate world came from the kitchens of mothers and grandmothers' in the United States. Most of these targeted the nuclear industry (Salleh 1997: 17). In the 1970s, a plethora of women's groups opposing uranium mining and nuclear power sprang up in North America, Australia, and Western Europe. In the United States, this opposition was galvanized by the meltdown at Three Mile Island, which prompted the first Women and Life on Earth conference in Amherst in March 1980 and the subsequent Women's Pentagon Action against nuclear war and weapon's development. In Britain, a sizeable Women for Life on Earth network was formed in 1982, when 30,000 women converged on the Greenham Common missile site, creating what became a permanent encampment and one of the most famous icons of ecofeminist resistance. In Australia, too, women played a prominent role in the Movement Against Uranium Mining and in the anti-nuclear and peace movements, and as Salleh recounts, feminists were to be found protesting at a number of key sites during the 1980s.[4] Opposition to nuclear power also grew in Eastern Europe following the Chernobyl nuclear reactor accident in 1986, which prompted concerned women across Eastern Europe, as well as in West Germany, to undertake a 'birth strike' (Salleh: 1997: 23).

Although Rachel Carson had already exposed the dangers inherent in the agricultural use of toxic chemicals in her book *Silent Spring* of 1962 (1965), it was not until the late 1970s that a more broadly based anti-toxics movement emerged among women, primarily in urban areas. The most famous story from this movement is doubtless that of Love Canal in the United States, which Merchant and Mellor both retell in some detail. This, Merchant comments, 'is a story of how lower-middle-class women who had never been environmental activists became politicized by the life-and-death issues directly affecting their children and their homes and succeeded in obtaining redress from the State of New York' (1996: 12).[5] Lois Gibbs, who initiated this campaign, subsequently went on to found the Citizen's Clearing House for Hazardous Waste. Throughout the world, as Merchant notes, 'the majority of activists in the grassroots movement against toxics are women' (1996: 12), and in the United States, many of these are now working-class, African American, Hispanic, and Native American women, whose neighbourhoods and reservations tend to be worst affected by this form of pollution (1996: 164). Native American women, especially through their association Women of All Red Nations, have also been prominent in opposing uranium mining, drawing attention to the high incidence of miscarriage, birth defects, and leukaemia on reservations contaminated by uranium tailings.[6] In Australia, too, uranium mining affects indigenous communities most immediately, and it is interesting to note that Aboriginal women, led by Yvonne Margarula of the Mirrar people, have become prominent in the present campaign against extension of the Ranger uranium mine to Jabiluka.

While the women's anti-toxics movement, and to some extent also the movement against uranium mining, have contributed to the growing awareness of the interstructuration of environmental degradation with classism and racism, as well as sexism, the ecological struggles of women in developing countries have fuelled a growing awareness of the unacceptable costs of western-style capitalist development. Reflecting the wider range of responsibilities carried by women in less industrialized societies in providing a subsistence base for their families, most of these struggles have been related to land-use practices. As early as 1964, for example, Brazilian women established the Acao Democratica Feminina Guacho, which, according to Salleh, 'soon evolved into an advocacy group for sustainable agriculture' (1997: 17). During the 1970s, the Chipko ('tree-hugging') movement was formed in Northern India to protect forests which had traditionally been used by village women as a source of food, fuel, fodder, and medicinal plants and were now being devastated by logging, while women in Kenya established the Greenbelt Movement to replant trees on degraded land. Concern about the ecologically and culturally destructive impact of inappropriate models of development grew during the 1980s, leading some women's groups to denounce the whole ideology of development, at least in its dominant modality, as a form of capitalist patriarchalism neo-colonization.[7] This line of critique has subsequently been developed further on a theoretical level by Vandana Shiva and Maria Mies, whose 'subsistence perspective', as I will indicate below, plays an important part in the ecofeminist arguments of Mellor and Salleh, although, as Mellor herself admits, this approach has been criticized for its alleged tendency to totalize and romanticize the situation of 'the valiant "Third World Woman"'.[8]

In 1992, the diverse, if generally not divergent, environmental concerns of women from all over the world were brought together under the auspices of the women's NGO meeting, dubbed the 'Planeta Femea', at the Earth Summit in Rio de Janeiro. This event is accorded considerable significance by all three authors.[9] As Mellor notes, 'the remarkable unity among the delegates […] led to a lack of sensitivity to divisions and inequalities between women' and, arguably, facilitated the ready cooptation of their proposals into 'what eventually became very watered down politics' (Mellor 1997: 36f).[10] Planeta Femea nonetheless disclosed the truly global dimensions of women's environmental activism in the early 1990s and provided the basis for further international ecofeminist networking in the following years, especially around the issue of genetic engineering, biotechnology, and 'bio-prospecting' of indigenous peoples and their lands.

By contrast with the geographical, socio-economic, and 'racial' diversity of women's grassroots environmentalism—not all forms of which, it should be noted, are explicitly feminist in orientation—the elaboration of ecofeminist theory has, until relatively recently, been dominated by the perspectives of white, middleclass women, above all from the English-speaking world. In her useful, if somewhat schematic, discussion of the various streams within western ecofeminism, Merchant distinguishes four main approaches—liberal, Marxist or socialist, cultural and social (i.e. anarchist)—which differ in their assumptions about 'nature' and 'human nature' (including the nature of sex/gender difference), in their critiques of other forms of environmentalism, in their understanding of what constitutes a feminist environmentalism, and in their vision for the future (1996: 5–18).[11] Interestingly, whereas both Mellor and Salleh see liberal feminism as incompatible with ecofeminism (Mellor 1997: 6; Salleh 1997: 114f), Merchant is more generous in acknowledging what women can achieve as environmentalists within a liberal feminist framework, which is to say, in pursuing equal educational and professional opportunities in order to work alongside men in meliorating environmental problems through better, more ecologically oriented science, technology, legislation, and—I would add, thinking in particular of the work of Karen Green—philosophy (1994).[12] Cultural and social(ist) ecofeminists, on the other hand, agree that rather than seeking equality in society as it is presently structured, women should be working towards a far more radical transformation of our worldview and social relations. Whereas the former tend to foreground female embodiment as the primary locus of connection between women and nature, seeking to initiate changes in consciousness through the development of new cultural practices, primarily of a spiritual and aesthetic kind, the latter focus more on women's socioeconomic position, drawing attention to the interstructuration of gender, class, 'race', and species domination, and calling for changes in the division of labour and distribution of wealth. Cultural ecofeminists share with Deep Ecologists an emphasis on personal psychological change in 'greening' the self and resacralizing the earth, just as social(ist) ecofeminists share with social ecologists and ecosocialists a critique of the capitalist mode of production. However, both strands of ecofeminism are critical of malestream green thinking, of whatever persuasion, to the extent that it fails to theorize the connections between the domination of women and nature.

It has become something of a commonplace practice in discussions of ecofeminism to distinguish between the 'essentialist' view of the woman-nature nexus allegedly adopted by cultural feminists and the 'constructivist' position of social(ist) and poststructuralist feminists. Merchant, too, suggests that what cultural ecofeminists seem to be saying is that 'what men do to the planet is bad; what women do is good' (1996: 13). As Mellor shows, however, this distinction is quite misleading. If 'essentialism' is taken to mean biological determinism—that is, the claim that attitudinal and behavioural differences between men and women are inherent in their natures as biologically male and female, such that women are naturally 'closer to nature' or more ecologically benign than men—then it should be stressed that very few, if any, ecofeminists make this claim. Andree Collard and Mary Daly certainly come very close to naturalizing women's alleged affinity with ecology, but, in the case of the former at least, it becomes clear that the problem lies not so much in any ontological difference between the sexes, but in the 'separatist mentality and dominating dualism of patriarchy' (Mellor 1997: 75).[13] Conversely, Mellor observes that Firestone's technosocialist prescriptions for escaping biology are premised on her assumption of a fundamental inequality between the sexes grounded in nature (1997: 81f). Meanwhile, poststructuralist feminists such as Luce Irigaray and Helene Cixous have tended to reify the man-culture, woman-nature dualism to such an extent that the only site of resistance to 'phallogocentrism' remaining to women appears to be the sexed and sensual female body (Mellor 1997: 99).[14] Susan Griffin's influential early work, *Women and Nature: The Roaring Inside Her* (1978), on the other hand, implicitly identifies the assimilation of women to nature as a culturally specific patriarchal construct. Mellor notes that Griffin's later work 'makes it clear that she adopts a social constructionist position' (1997: 49) (as do the prominent spiritual ecofeminists Starhawk and Charlene Spretnak), but as Gloria Feman Orenstein has observed, *Women and Nature* was itself 'an early critique of essentialism which historicizes and contextualizes in detail the parallel oppression of women and nature' (1990: 20).[15]

Like most cultural ecofeminists' texts, however, Griffin's was a work of not only critique but also affirmation. This is why it has been so empowering for some and so problematic for others. Having shared a common history of oppression with nature, in the course of which women were largely excluded from the western male project of mastery and transcendence, thereby remaining closer to the realm of embodiment and immanence, 'we' were now, Griffin suggests, in a privileged position to *speak for* nature in the context of pursuing 'our' own social emancipation. Reversing the common evaluation of the woman-nature connection, Griffin effectively redefines the feminist project as an emancipation *with* nature, rather than *from* it, as in the dominant Enlightenment tradition. This was a truly radical move, and one that remains fundamental to any deeper ecofeminist transformation of culture and society. However, the early cultural feminist celebration of women's alleged closeness to nature, while not necessarily biologically determinist, certainly appears problematic from a contemporary perspective. Firstly, there is a tendency in some cultural feminist writing to simply reverse the value judgements attaching to the dualistically opposed terms of Man/Woman, Culture/Nature, Reason/Emotion, and Mind/Body, without adequately questioning the content of these terms or the structure of dualism itself. Plumwood refers to this approach as the 'feminism of uncritical reversal' (1993: 31). To the extent that patriarchal constructions of the feminine are not adequately distinguished from the actual historical experience of women, they tend to return to haunt the ecofeminist vision of a liberated 'female nature'. This is possibly exacerbated by the highly poetic style of much early cultural ecofeminist writing, including Griffin's and Daly's. It may well be, as Carol Christ has argued, that the 'revolution in thought' embodied by the new ecofeminist paradigm requires new modes of writing (1990: 62). But it is important that salient distinctions and critical insights do not get lost along the way.

Among the distinctions that have frequently been overlooked in cultural ecofeminism are those separating women in different times and places, as well as in terms of class and 'race'. Generalizing from the experience of some women, interpreted in a certain way, to the experience of all women, many early ecofeminists were certainly prone to 'essentialism' in the secondary sense of a tendency to universalize. Often underlying this universalization of 'women's experience' is the radical feminist prioritization of the patriarchal oppression of women as the root cause of all other structures of domination. As Mellor observes, this kind of reductionism has compounded the view of ecofeminism as 'essentialist' (1997: 96). Many cultural feminists have sought support for this prioritization of sexism in research on that epochal shift which appears to have taken place in the Mediterranean region around 3000 BCE from a matrilineal, matrifocal, seemingly peaceful, relatively egalitarian, earth goddess-worshipping culture to a patriarchal warrior society, characterized by new tools, weapons, hierarchies, and sky gods.[16] As ecofeminist theologian Rosemary Radford Ruether has observed, cultural feminists have tended to interpret the evidence for this transition on the mythical model of the Fall, whereby the subjection of women to male domination, often said to have been brought about by invading Indo-Germanic horsemen, is seen to have led simultaneously to the masculinization of culture and the institution of a more aggressive and exploitative relationship to the land and to other peoples (1992: 147ff). The notion of a Fall into patriarchy has doubtlessly been valuable as an enabling myth for many women, holding out the hope that since things were different once, they can be so again. However, as Mellor observes, this theory fails to explain how patriarchy emerged among the war-mongering Kurgan, as well as harbouring potentially racist assumptions in attributing the 'disease' of patriarchy to the invading 'stranger' (1997: 152). It has, moreover, also generated a certain kind of feminist 'bad faith', especially where it is implied that women have retained certain psychological or cultural links to this era of primordial harmony with nature that have been lost to men. Suppressing complicity, some cultural ecofeminists have been reluctant to confront the extent to which women, too, have been deformed by patriarchy. Prioritizing sexism, they have also tended to overlook significant socio-economic factors underlying the current ecological crisis and intersecting with patriarchal structures of domination in ways that do not impact upon all women in the same manner or to the same degree.

Within both social and socialist ecofeminism, these factors generally receive considerably more attention. Social ecofeminism, according to

Merchant, 'envisions the restructuring of society in humane, decentralized communities', and draws on the work of eco-anarchist Murray Bookchin. Apart from Janet Biehl, who has now renounced any affiliation with ecofeminism, Merchant identifies Salleh, along with Val Plumwood, Ynestra King, and Chaia Heller, as social ecofeminist (Merchant 1996: 13–14).[17] Mellor, on the other hand, who goes into the complexities of the latterly rather fraught relationship between social ecology and ecofeminism in considerably more detail, refers to Salleh as a socialist ecofeminist (1997: 139).[18] I would tend to agree with Mellor's judgement here, to the extent that Salleh certainly does not align herself with Bookchin. However, Merchant is doubtless correct in wanting to differentiate Salleh's position from her own more reformist democratic socialism, insofar as Salleh's work—like that of Plumwood, King, Heller, and indeed Mellor—appears to prefigure a considerably more radical overcoming of modernity than does her own.

'Socialist ecofeminism', observes Merchant, 'is not yet a movement, but rather a feminist transformation of socialist ecology that makes the category of reproduction, rather than production, central to the concept of a just, sustainable world'. Like cultural ecofeminism and Deep Ecology, it assumes that 'nature is an active subject, not a passive object to be dominated,' but it also goes beyond both 'in offering a critique of capitalist patriarchy that focuses on the dialectical relationships between production and reproduction, and between production and ecology' (Merchant 1996: 15). Merchant's own valuable contribution to this critique has been above all as a historian investigating those interrelated transformations in the position of women, the view of nature, and the mode of production.

As becomes apparent from the essays in the first section of *earthcare*, Merchant's thinking has been influenced by poststructuralism in recent years. This is particularly evident in the essay on 'Eve: Nature and Narrative', a longer version of which was first published in 1995,[19] in which Merchant examines the role of cultural paradigms in shaping prevailing views of women and nature, in this case, the Biblical narrative of the Fall. From the seventeenth century, this narrative model of loss/decline/corruption followed by recovery/restoration, Merchant writes, came to underwrite the project of scientific discovery, colonial conquest, and *laissez-faire* capitalism. She suggests we need a new narrative which would allow for a greater degree of complexity and uncertainty, and incorporate a greater variety of voices, especially of those who have hitherto been silenced, a new narrative, moreover, that would ultimately need to be written 'through action' (Merchant 1996: 54–56). In the following essay on women and science, it becomes clear that this new paradigm also implies a new history, theory, and practice of science: a history which acknowledges the contributions of non-European and non-male scientists; a theory which—following Donna Haraway—recognizes the epistemological limits or 'situatedness' of all scientific research, while nonetheless resisting the radical constructivist denial of the possibility of any materially grounded knowledge of reality; and a practice which is respectful of the independent agency of nonhuman nature and oriented towards partnership rather than domination (Merchant 1996: 57–72; Haraway 1988: 183–201).

The 'partnership ethic' which Merchant outlines in more detail in the concluding chapter is her preferred model to the 'ethic of care' advocated by many ecofeminists. Deane Curtin (1991), for example, drawing on the work of Nancy Chodorow and Carol Gilligan, has argued that women's apparently more relational sense of self, their greater concern with the concrete avoidance of harm rather than with abstract rights and rules, could provide the basis for a new ecological ethic of care. The danger with this approach, in Merchant's view, is that it could be seen to reinforce the ideological assumption that 'women's nature is to nurture' (1996: 8). As Val Plumwood has observed, it also creates a distorted understanding of our relationship with non-human others, failing as it does to acknowledge their independent agency and the fact that their interests will not necessarily coincide with ours. An ethic of care will not get you very far, if, as once happened to Val, you are being attacked by a very large crocodile! (1993: 156–159). Not unlike Plumwood's 'ethic of mutuality', Merchant's model of partnership avoids the hubris of constructing nature either as a domain to be dominated or as a patient to be healed, acknowledging both our dependence upon a fundamentally 'disorderly order' and the responsibility for carefully considered ecological action that we have nonetheless acquired along with the technological power to destroy life as we know it today (1996: 217–220). Such a partnership ethic, moreover, conjoins respect for non-human nature as an 'equal subject' with a homocentric concern for social justice in the equitable and cooperative fulfilment of all people's 'vital human needs'. Acknowledging cultural diversity, as well as biodiversity, Merchant's model of partnership among humans and with nature is finally also a situational, rather than a universalist ethics: individual ethical and policy decisions, she writes, will need to be 'negotiated by a human community in a particular place, but the outcome will depend on the history of people and nature in the area, the narratives they tell themselves about the land, vital human needs, past and present land-use patterns, the global context, and the ability or lack of it to predict nature's events' (1996: 221f).

Mellor's *Feminism and Ecology* is exemplary in its clarity of style and structuration. Her analysis moves in logical stages through a general discussion of women and the environment to an overview of ecofeminist thought, before tackling in her two central chapters the vexed issues of 'essentialism' in theorizing women's embodiment and of women's 'privileged standpoint' with regard to ecology. She then proceeds to position ecofeminism in relation, firstly, to the Green movement, especially Deep Ecology, and, secondly, to social ecology and ecosocialism, before summarizing her conclusions concerning the 'material connection' between feminism and ecology. Salleh, on the other hand, elaborates her argument elliptically, through a series of dizzying leaps and swerves, in a manner as postmodern as her model of ecofeminist politics. By contrast with Mellor's measured academic prose, unruffled by emotion until at last in the final paragraph she rather self-consciously allows herself 'a cheer',[20] Salleh's impassioned treatise is clearly carried by rage and grief, as well as by a certain irrepressible optimism, and, underneath it all, a deep delight in 'all that walks and flies and swims and stands' (Salleh 1997: xv). Whereas Mellor may well succeed in winning over some sceptics, Salleh is bound to unsettle, annoy, and even outrage many readers. It is nonetheless to be hoped that this book, no less than Mellor's and Merchant's, will not just speak to the converted.

Despite these evident differences in style and approach, Mellor and Salleh share a similar point of departure in theorizing the connections between women and nature, feminism and ecology, in that both build on the work of Maria Mies and Vandana Shiva. Mellor and Salleh credit

Mies (1986) with developing the first substantial materialist ecofeminist critique of capitalist patriarchy in arguing that the accumulation of capital through the creation of surplus value has always been dependent upon the exploitation of the largely unpaid labour of women as housewives, and of colonized peoples as subsistence farmers and low-paid piece workers. In making this argument, Mies effectively deconstructed the opposition of production and reproduction by maintaining that gestating, birthing, suckling, and caring for a young child were no less 'truly human', no less a 'conscious social activity' than any paid work performed by means of the head and hands. The extraction of surplus value from a predominantly male paid workforce has only been possible by tapping into women's domestic labour in the production of life, and the means of life, just as the global expansion of capitalism has been premised on the subsistence labour of colonized peoples, especially, as Vandana Shiva has argued, on women, whose livelihood it also threatens through environmental degradation and social dislocation (Mellor 1997: 169–171; Salleh 1997: 60–62; Shiva 1989).

Salleh argues that this perspective on women's work necessitates a critical rethinking of certain key aspects of Marxist theory. Firstly, while she denies that Marx subscribed to a 'crude domination ethic' with regard to non-human nature, Salleh observes that his labour theory of value was both anthropocentric and androcentric in failing to factor in the productivity of the land and of women's bodies, as well as in its implicit disdain for the subsistence and domestic labour of peasants and housewives (1997: 70–74). Indeed, Salleh argues that if women as unpaid domestic and subsistence workers are collectively the group at once most exploited and most marginalized by global capitalism throughout the world, the group with the most 'radical chains' and with certain ecologically valuable alternative values and attitudes, then it is they who are now poised to emerge as the hidden subject of History and Nature, bringing forward a revolutionary transformation conjoining feminism, ecology, socialism, and postcolonialism (Salleh 1997: 3–14, 190–193).

Mellor is far more circumspect in her claims for ecofeminism, insisting that to 'start from women's experiences is not to claim centrality or priority over other oppressions' (1997: 175). She does argue nonetheless that the analysis of women's socio-economic position provides the basis for a more fundamental critique of both patriarchy and capitalism than that of either non-ecological feminism or non-feminist ecosocialism (1997: 77, 174). Moreover, she is in agreement with Salleh in maintaining that the nature of women's work potentially gives them a relatively privileged standpoint in relation to ecology. Labour in the *production of life*, Mellor observes, is work that in effect mediates between nature and culture. As Sherry Ortner argued in 1974, it was primarily because of their work in mediating nature for men that women had been subordinated to men in the context of a culture oriented towards the transcendence of nature (1974: 67–87). As social(ist) ecofeminist Ynestra King has more pointedly restated this argument:

> It is as if women were entrusted with and have kept the dirty little secret that humanity emerges from non-human nature into society in the life of the species and the person. The process of nurturing an unsocialized, undifferentiated human infant into an adult person—the socialization of the organic—is the bridge between nature and culture. The western male bourgeois then extracts himself from the realm of the organic to become a public citizen, as if born from the head of Zeus. (1989: 116)

Ortner's solution to the problem of women's subordination was, like de Beauvoir's, to advocate that they join men in the creation of culture. From an ecological perspective, however, this is too undialectical. For it is precisely the urge to escape from nature, by means of a form of culture premised on the denial of dependence, which is the problem. This, Mellor observes in her conclusion, is the culture of the 'filiarchs', the ruling oligarchy of 'sons', who, lacking even the sense of responsibility that one might associate with patriarchal rule, continue to '"play" in the world of transcendence', blind to the ecological and social costs of their privilege. While filiarchs have existed in other times and cultures, the currently most destructive and oppressive form of filiarchy globally is that of the new capitalist world 'order' (1997: 193f). With reference to what ecosocialist Martin O'Connor terms the 'parasitism' of the most affluent two-thirds of today's richer societies, Mellor writes that a 'minority of the human race is able to live as if it were not embodied and embedded, as if it had no limits, because these limits are borne by others, including the earth itself' (1997: 190).

Rather than striving for an ecologically unsustainable level of 'transcendence' for all, Mellor argues, we need a philosophy and a politics that revalues 'immanence'. Such a philosophy and politics of immanence would need to draw on the experiences and perspectives of those who have disproportionately carried the burden of human embodiment and embeddedness. It is in this connection that Mellor, like Salleh, believes ecofeminists are justified in attributing to women a certain 'epistemic advantage'. To the extent, that is, that most of the work associated with the production of life and the means of life still generally falls to women (whatever else they might do)—and this is evidently almost as true in postfeminist Australia and post-Communist Russia as it is in India or Nigeria—they are likely to have a greater awareness of the nature and consequences of human embodiment than do those who are able to leave the labour of mediation to others. It is presumably for this reason that women—whether or not they have seen themselves as 'feminist'—have been at the forefront of protest movements in many parts of the world, highlighting contradictions between the creation of 'wealth' or the protection of 'freedom' and the continuation of life. Women's domestic and subsistence work, as well as to some extent their work in the 'caring professions' can, moreover, be seen to provide a model for a more embodied epistemology and a more ecological praxis. Labouring in reciprocity with nature, as Salleh puts it, women—and other mediators—are more attuned to 'biological' (Mellor) or 'enduring' time (Salleh). This is the time of the human body, of the daily round of meeting its physical needs, which emerge out of longer cycles of health and sickness, growth and ageing, birth and death; and, especially in the case of subsistence farmers and indigenous people who are still leading a more traditional existence, the cyclical time of plant and animal life and seasonal change (Mellor 1997: 172). This kind of work, Salleh suggests, engenders a 'kinaesthetic' way of knowing, which contrasts with the predominantly specular logic of patriarchal reason and suggests the possibility of a more embodied and non-dualistic materialism. Women's caring labours also provide a model for a non-instrumental ethic, which, following Sara Ruddick, Salleh terms a 'holding ethic', one that is oriented to maintaining interconnectedness, acknowledging vulnerability, allowing for replenishment, and minimizing conflict (Salleh 1997: 138–147;

Ruddick 1989). Not dissimilarly, Mellor argues that the analysis of women's experiences in the mediation of nature discloses a particular kind of subaltern 'situated knowledge': that, namely, of an 'immanent realism', as 'revealed through patterns of subjugation and the perspectives they generate within the human community, and through an awareness of the interrelatedness of humanity and nature in ecological processes' (1997: 111).

One of the key challenges for ecofeminism, in Mellor's view, is 'to address the central question of how to theorize the finite nature of the planet and the biological differences between men and women without falling into ecological and biological determinism' (1997: 166). Mellor herself rises to this challenge in a very convincing manner. As she observes at the outset, the 'postmodern/poststructuralist domination of contemporary social theorizing is presenting us with a false choice between radical social constructivism and various forms of universalism and essentialism' (1997: 7). With reference to the work of Diana Fuss, she points out that this kind of radical constructivism is itself essentialist to the extent that it is universalist and reductivist in its claims for the absolute priority of culture over nature (Mellor 1997: 97). Moreover, in reducing physical reality to a passive screen for our culturally encoded and psychosocially determined projections, radical constructivism is 'as human-centred and arrogant as the Enlightenment science that Merchant condemned', as well as radicalizing the flight from embodiment and embeddedness that has been so central to the modernist project (Mellor 1997: 124; Salleh 1997: 106). Instead, Mellor argues for a genuinely postmodern 'ecological holism' which would recognize that all human existence, however socially mediated and culturally framed, grows out of and is enfolded by natural processes that are 'material, real, dynamic, and always beyond human knowing', as well as radically uncertain in their outcome (1997: 185). Among such natural processes, Mellor includes those which generate human sexual difference. Contra Judith Butler, she holds to the view that men and women are differently embodied and that this difference has had certain consequences for the positioning of men and women in relation to the labour of mediation (1997: 9). Mellor nonetheless resists biological determinism by insisting that this does not mean that their natures are any more 'fixed' than is nature generally, and by acknowledging the role of social constructions in the interpretation of sexual differences along gender lines (1997: 97, 174).

While thus acknowledging sexual difference as pertinent to an ecofeminist critique, Mellor is careful to reject the view that women's epistemic advantage arises directly from their embodiment as female or that it is exclusive to women. Throughout her book she stresses that subordinate men, too, are in various ways made to bear the burden of mediation, while privileged women are among their beneficiaries (Mellor 1997: 196). For Mellor, ecofeminism is but one element in a wider politics of ecological and social transformation which is to be built out of a coalition of many groups and peoples throughout the world who are in various ways, and to different degrees, exploited and marginalized by the prevailing structures of parasitical transcendence (Mellor 1997: 192). Similarly, Salleh writes of the necessity of forging alliances, especially between ecofeminism and indigenous movements, and she frequently points to connections between the position of women and other colonized subjects, again particularly indigenous people, while nonetheless recognizing the complicity of more privileged women in structures of exploitation.

Ecofeminist politics, as Salleh astutely observes, is 'a transitional praxis by historically contingent subjects' (1997: 192). To the extent that ecofeminists do not seek to shore up the conventional sex/gender division of labour, or to reaffirm traditional gender dualisms—and the overwhelming majority do not—their project is a profoundly dialectical one. As Val Plumwood puts it, it is to 'critically affirm' women's historically and socially constituted different relation to nature as a source of alternative knowledges and practices in the context of ecological transformation, while simultaneously working, as feminists, to dismantle those structures of domination in which this difference is largely grounded. We are called to walk a tightrope, constantly on our guard against falling one way—into an uncritical identification with a revalued, but still dualistically defined 'femininity' and 'nature'—or the other—into an equally dualistic disavowal of embodiment and embeddedness. Critically affirming women's difference, without overlooking women's differences, involves recognizing 'female identity', understood historically rather than ontologically, as 'an important if problematic tradition which requires critical reconstruction, a potential source of strength as well as a problem, and a ground of both continuity and difference with traditional ideas' (Plumwood 1993: 64). Ecofeminist critique is not about naively celebrating women's proclaimed corporeality, connectedness, and closeness to nature. For the aim of ecofeminist transformation is not to reify existing gender differences, but rather to facilitate the relocation of women (and other subordinated groups) into culture as well as nature, and of men into nature as well as culture, while simultaneously contributing to the redefinition and mutual reattunement of culture and nature. To the extent that ecofeminism succeeds in this tricky tightrope act, it can indeed claim to be, as Plumwood and Salleh put it, a 'third wave' of feminism, dialectically sublating the opposition between the feminism of equality and that of difference, between socialist and cultural feminism (Plumwood 1993: 39; Salleh 1997: 104). Merchant, Mellor, and Salleh all exemplify this move, not only in their critical affirmation of women's perspectives and practices in the cultural mediation of nature, but also in their acknowledgement of the potential value of insights gained through spiritual practices, alongside the necessity for fundamental change at the level of our socio-economic relations of production and (re)production.[21] In this respect, ecofeminist thinking can also be seen to be moving beyond the old Marxist base-superstructure dualism in a way which holds out the promise of overcoming the unfortunate and unnecessary division between various forms of deep and social(ist) ecology (Gottlieb 1996).

In conclusion then, a final word on means and ends. Both Mellor and Salleh look towards a future society in which 'sustainability and social equity can go together' (Salleh 1997: 180), where responsibility for attending to the consequences of human embodiment and embeddedness would be shared, and where production would be cooperative, orientated to need rather than greed, and low in its environmental impact (Mellor 1997: 196). This would indeed represent a radical transformation of the prevailing order of things, especially for the industrialized world. I too find this vision highly appealing. The problem, of course, is how to get from here to there. In the absence of any guarantee on the future, moreover, it becomes all the more important that the qualities of the desired end come to characterize the chosen means. In this respect too, namely as a mode of activism that is typically non-violent and cooperative, seeking alliances across difference, rather than anxiously sectarian, and embodying an integrative, corporeal, sometimes soul-full and often playful form of reason, ecofeminist politics might also be seen as exemplary.

Notes

1. For a translation, see d'Eaubonne 1994.
2. Salleh, for example, refers to three specifically lesbian-identified anti-nuclear groups: Lesbians United in Non-Nuclear Action (against the Seabrook reactor), Dykes Opposed to Nuclear Energy, who 'organized a New York conference on the energy crisis as a malegenerated pseudo-problem', and Dykes Against Nukes Concerned with Energy (against the United Energy) (Salleh 1997: 19–20).
3. Salleh herself has been an energetic activist on a number of fronts: in the Movement Against Uranium Mining, the Franklin Dam Blockade, the Australian Greens, the Society for Social Responsibility in Engineering, the Women in Science Enquiry Network, and the Women's Environmental Education Centre in Sydney, as well as in some other more localized campaigns.
4. For example, outside the Smithfield air force base in South Australia, the Lucas Heights Atomic Energy Establishment in NSW, and, together with Aboriginal men and women and other peace activists, at the US reconnaissance station at Pine Gap (Salleh 1997: 18–22).
5. In 1978, alarmed by the high incidence of miscarriage, birth defects, and unusual and potentially life-threatening health problems affecting women and children in their neighbourhood, the women of Love Canal, led by Lois Gibbs, initiated investigations which revealed that the State of New York had given approval for the development of their residential area near a site that the Hooker Chemicals and Plastics Association had used as a toxic waste dump between 1942 and 1953. Tragically, the elementary school had been built right over the dump itself (See Merchant 1996: 11–12, 155–157; Mellor 1997: 20–22).
6. WARN was founded in South Dakota in 1977 to protest against involuntary sterilization, the erosion of the family on reservation lands, and the shrinkage of reservation lands (see Merchant 1996: 155).
7. For example, Sen and Grown 1987. This is the report presented to the 1985 UN Decade for Women meeting in Nairobi by Development

Alternatives with Women for a New Era (DAWN), a group of 22 activists, researchers, and policy makers from Africa, Asia, and Latin America (see also Mellor 1997: 30–33).

8. For example, in Agarwal 1992; Braidotti 1994; Mellor 1997: 35.
9. Merchant's concluding chapter is in fact a revised version of the address she gave to the Planeta Femea conference (Merchant 1996: 209–224). See also Mellor 1997: 35–37 and Salleh 1997: 26–27, 136.
10. Salleh is even more damning of the final document, commenting that 'the Rio meeting provided a template for the neo-feudal order and its key stratifications' (1997: 135).
11. There is a discrepancy between Merchant's Table 2 on p. 6, which distinguishes Marxist and socialist ecofeminism, and her subsequent discussion, in which Marxist ecofeminism is apparently subsumed by socialist ecofeminism, which is now distinguished from social ecofeminism.
12. See also Green 1995.
13. With reference to Collard 1988. See also Daly 1978.
14. With reference to Irigaray 1985. Cixous is not included in Mellor's bibliography, but she is probably referring to her influential essay 'The Laugh of the Medusa' (1976). It should be pointed out that Irigaray has become more explicitly ecofeminist in her more recent work, and is apparently an active environmentalist. See Irigaray 1993.
15. See also Starhawk 1990. and Spretnak 1990.
16. See, for example, Gimbutas 1982.
17. Bookchin has developed his theory of social ecology through a number or publications from *Toward an Ecological Society* (1980) to *Re-Enchanting Humanity* (1995). Janet Biehl broke with ecofeminism in her book *Finding Our Way: Rethinking Ecofeminist* (1991). Ynestra King began in the social ecology movement, giving lectures at the Institute of Social Ecology in Vermont, founded by Murray Bookchin, but has since distanced herself from his approach. As well as being a prominent activist, she has published a number of highly significant articles since the early 1980s. See, for example King 1989 and 1990.
18. For her excellent discussion of social ecology and ecofeminism, see pp. 150–161.
19. See Merchant 1995.
20. Having realized that the acronym of her key terms—Holism, Uncertainty, Responsibility, Reciprocity, Awareness, and Humility—spelt HURRAH!

21. Merchant is less explicit about this, but agrees that practices orientated towards facilitating change at the level of consciousness are no less important than those of a more conventional political nature.

Bibliography

Agarwal, Bina. 1992. The Gender and Environment Debate: Lessons from India. *Feminist Studies* 18 (1): 119–158.

de Beauvoir, Simone. 1968. *The Second Sex*. London: Jonathan Cape.

Beinssen-Hesse, Silke, and Kate Rigby. 1996. *Out of the Shadows: Contemporary German Feminism*. Melbourne: Melbourne University Press.

Biehl, Janet. 1991. *Finding Our Way: Rethinking Ecofeminist Politics*. Montreal: Black Rose Books.

Bookchin, Murray. 1980. *Toward an Ecological Society*. Montreal: Black Rose Books.

———. 1995. *Re-Enchanting Humanity*. London: Cassell.

Braidotti, Rosi, et al., eds. 1994. *Women, the Environment and Sustainable Development*. London: Zed Books.

Carson, Rachel. 1965. *Silent Spring*. London: Penguin.

Christ, Carol P. 1990. Rethinking Theology and Nature. In *Reweaving the World: The Emergence of Ecofeminism*, ed. I. Diamond and G.F. Orenstein. San Francisco: Sierra Club Books.

Cixous, Hélène. 1976. The Laugh of the Medusa. *Signs* 1: 245–264.

Collard, Andree (with Joyce Contrucci). 1988. *Rape of the Wild*. London: The Women's Press.

Curtin, Deane. 1991. Toward an Ecological Ethic of Care. *Hypatia: Special Issue on Ecological Feminism* 6 (1, Spring): 60–74.

Daly, Mary. 1978. *Gyn/Ecology: The Metaethics of Radical Feminism*. London: The Women's Press.

d'Eaubonne, Françoise. 1974. *Le Féminisme ou la mort*. Paris: Pierre Horay.

———. 1994. The Time for Ecofeminism. In *Ecology*, ed. C. Merchant and Trans. R. Hottell, 174–197. Atlantic Highlands, NJ: Humanities Press.

Firestone, Shulamith. 1970. *The Dialectic of Sex*. London: The Women's Press.

Gimbutas, Marija. 1982. *The Goddesses and Gods of Old Europe*. Berkeley: University of California Press.

Gottleib, Roger S. 1996. Spiritual Deep Ecology and the Left: An Attempt at Reconciliation. In *The Sacred Earth, Religion, Nature and Environment*, ed. R.S. Gottlieb, 516–531. New York: Routledge.

Green, Karen. 1994. Freud, Wollstonecraft, and Ecofeminism: A Defence of Liberal Feminism. *Environmental Ethics* 16 (Summer): 117–134.

———. 1995. *The Woman of Reason: Feminism, Humanism and Political Thought*. Cambridge: Polity Press.

Griffin, Susan. 1978. *Women and Nature: The Roaring Inside Her*. New York: Harper and Row.

Haraway, Donna. 1988. *Simians, Cyborgs, and Women: The Reinvention of Nature*. New York: Routledge.

Irigaray, Luce. 1985. *Speculum of the Other Woman*. Trans. G.C. Gill. Ithaca: Cornell University Press.

———. 1993. *Je, te, nous. Toward a Culture of Difference*. Trans. A. Martin. London and New York: Routledge.

King, Ynestra. 1989. Toward an Ecological Feminism and a Feminist Ecology. In *Healing the Wounds: Promise of Ecofeminism*, ed. J. Plant, 18–28. Santa Cruz: New Society Publishers.

———. 1990. Healing the Wounds: Feminism, Ecology, and the Nature/Culture Dualism. In *Reweaving the World: The Emergence of Ecofeminism*, ed. I. Diamond and G.F. Orenstein. San Francisco: Sierra Club Books.

Mathews, Freya. 1994. Relating to Nature: Deep Ecology or Ecofeminism? *Trumpeter* 11 (4): 159–172.

Mellor, Mary. 1997. *Feminism and Ecology*. Cambridge: Polity Press.

Merchant, Carolyn. 1985. Preserving the Earth. Women and the Progressive Conservation Crusade. *Environmental Review* 8 (I): 57–85.

———. 1995. Reinventing Eden: Western Culture as Recovery Narrative. In *Uncommon Ground: Toward Reinventing Nature*, ed. W. Cronon. New York: Norton.

———. 1996. *Earthcare: Women and the Environment*. New York: Routledge.

Mies, Maria. 1986. *Patriarchy and Accumulation on a World Scale*. London: Zed Books.

Orenstein, Gloria Feman. 1990. The Greening of Gaia: Ecofeminist Artists Revisit the Garden. In *Reweaving the World: The Emergence of Ecofeminism*, ed. I. Diamond and G.F. Orenstein. San Francisco: Sierra Club Books.

Ortner, Sherry B. 1974. Is Female to Male as Nature is to Culture? In *Woman, Culture, Society*, ed. M.Z. Rosaldo and L. Lamphere, 67–87. Stanford: Stanford University Press.

Plumwood, Val. 1993. *Feminism and the Mastery of Nature*. London: Routledge.
Ruddick, Sara. 1989. *Maternal Thinking*. London: The Women's Press.
Ruether, Rosemary Radford. 1992. *Gaia and God: An Ecofeminist Theology of Earth Healing*. San Francisco: Harper & Collins.
Salleh, Ariel. 1997. *Ecofeminism as Politics: Nature, Marx and the Postmodern*. London: Zed Books.
Sen, Gita, and Caren Grown. 1987. *Development, Crises and Alternative Visions*. New York: Monthly Review Press.
Shiva, Vandana. 1989. *Staying Alive: Women, Ecology and Development*. London: Zed Books.
Spretnak, Charlene. 1990. Ecofeminism: Our Roots and Flowering. In *Reweaving the World: The Emergence of Ecofeminism*, ed. I. Diamond and G.F. Orenstein. San Francisco: Sierra Club Books.
Starhawk. 1990. Power, Authority and Mystery: Ecofeminism and Earth-Based Spirituality. In *Reweaving the World: The Emergence of Ecofeminism*, ed. I. Diamond and G.F. Orenstein. San Francisco: Sierra Club Books.

Kate Rigby started out as an academic at Monash University in German Studies and Comparative Literature, and has since been instrumental in the development of the emerging field of the Environmental Humanities. She is currently Professor of Environmental Humanities at Bath Spa University and Adjunct Professor of Literary Studies at Monash University. Her expertise within the Environmental Humanities are primarily in environmental literary studies, along with ecophilosophy, environmental history, ecology and religion. Together with Freya Mathews and Sharron Pfueller, she co-founded the journal *Philosophy Activism Nature* and was the founding president of the Association for the Study of Literature, Environment and Culture—Australia-New Zealand, and founding director of the Australia-Pacific Forum on Religion and Ecology. Her most recent monograph is *Dancing with Disaster* (2015).

5

Women and Land Claims

Deborah Bird Rose

In the Northern Territory (NT), the Aboriginal Land Rights (Northern Territory) Act 1976 (hereafter referred to as the Act) has had ambivalent effects. On the one hand, it has enabled more than 36% of the land and 86% of the coastline to be transferred to Aboriginal freehold title, and has thus enabled thousands of Aboriginal people to achieve a great measure of economic and political opportunity. On the other hand, the Act has created inequalities among Aboriginal people. The most publicly compelling type of inequality is derived from the fact that under the Act only unalienated crown land is available for claim, while land that was held as reserve land at the time of the passage of the Act became Aboriginal freehold land without having to go through the claim process. As a result, some groups of Aboriginal people are in legal possession of the whole or substantial parts of the country with which they assert a relationship of

Originally published as: Rose, Deborah Bird. 1995. Women and Land Claims. In *Land, Rights, Laws: Issues of Native Title*, 1–8. Canberra: Native Title Research Unit, Australian Institute of Aboriginal and Torres Strait Islander Studies.

D.B. Rose (✉)
University of New South Wales, Sydney, NSW, Australia

L. Stevens et al. (eds.), *Feminist Ecologies*,
https://doi.org/10.1007/978-3-319-64385-4_5

ownership or belonging. Other groups are in possession of portions of land so small as to utterly trivialize their aspirations. A few people are in possession of nothing at all under Aboriginal freehold title. Thus, some groups have been massively advantaged economically, culturally, psychologically and in terms of their long-range prospects for cultural and social survival.

By contrast, gender inequality has been pervasive throughout the history of claims to land, but has received far less public attention. Land claims until recently have involved a massive privileging of senior Aboriginal men vis-à-vis senior Aboriginal women. In this chapter, I consider some of the ways in which Aboriginal women have been disadvantaged by the privileging of men in a system that is predominantly controlled by men. I then discuss attempts Aboriginal women and their anthropologists and legal counsel have made to get more of their evidence into the land claim process. My urgent intention is to alert claimants, anthropologists and lawyers who are preparing Native Title cases to some of the precedents in the Act. The marginalization and exclusion that NT women have experienced are in clear contradiction to the intention of the Act, and must not be repeated in other parts of Australia under the more recent Native Title legislation.

Invisible Women

The Act brought into existence a public record consisting of much of the written materials prepared in advance of the hearing, the transcript of the public portions of the hearing, references to evidence and performance in the context of secret/sacred knowledge, and the final report by the Aboriginal Land Commissioner. This written record can be understood to document a people's relationship to land at the time of the inquiry, but the record is extremely narrow and gives a highly biased representation of Aboriginal women as landowners and as managers of country, of kinship and other social relations and of ecological, geographical, religious and other forms of knowledge. The spiritual dimension of their lives sometimes is not even mentioned.

The written record reflects processes of consultation, investigation, preparation, presentation and representation. It clearly reflects the male dominance of the legal profession and the greater numbers of men who have been employed as senior anthropologists in the preparation of land claims. The written record thus tends to confirm the androcentric heritage of anthropology as well as to reinforce the stereotype, commonly held by many men and women of non-Aboriginal culture, that Aboriginal societies are male dominated and that women are essentially pawns in social life. In an astonishing number of claims, it has been seen to be quite adequate for men to speak for women and for women to say virtually nothing on their own behalf[1].

The written record of land claims also stands as testimony to a tunnel vision approach on the part of land councils which asserts that as long as people get their land, it does not matter who gives evidence. In this view, gender equity is seen to be an optional extra that land councils simply cannot afford.

The tunnel vision approach depends on a view, which rarely is articulated (but which most people involved in land claims have heard at one time or another), that the Act has no bearing on Aboriginal people and their Law in the further course of their lives. The idea is that people present their case, get their land and get on with their lives. This simplistic view obscures the fact that a land claim is a process which can take up years of their lives, involve them in intense politicking, engage their deepest spiritual, emotional and intellectual endeavours, and radically change the conditions not only of their own lives but of the lives of their descendants as well.

The simplistic view is false as well as self-serving. Emerging from an assumption that a land claim is an alien procedure which is imposed upon Aboriginal people, this view would have us suppose that Aboriginal people do not seize the claim procedure itself. In fact, however, many land claims are treated by the claimants as a ceremony for land,[2] and in ceremony, the right people should be involved in the right ways. Land claims which exclude women as participants have a socially disruptive potential equivalent to that of ceremony improperly conducted.

In a successful claim, Australian law recognizes the authority and integrity of the claimants' Law by granting the land. One result is that a set of Law persons has been empowered, and their empowerment feeds back into the ongoing life of country and community. The people who spoke to the judge as Law persons and were subsequently found to be traditional owners within the terms of the Act are positioned very powerfully within their own local political systems. In Aboriginal societies, ownership of knowledge is translated into social power through making things happen, whether it be ecological, social, intellectual or spiritual. Getting land back is a superb example of the power not only of Law but also of the person who holds it and demonstrates it. Men have been massively advantaged economically, culturally, psychologically and in terms of their long-range prospects for political action.

The disadvantage for women is not only, or even predominantly, in matters of secret/sacred knowledge. Rather, the disadvantage for many women (not all) encompasses the full dimension of their right to speak with knowledge and passion about their status as landowners. If the anthropologists work most closely with men (as most male anthropologists are encouraged to do), and if the lawyers work most closely with men (and all the lawyers who have had the responsibility of carrying a land claim have, to date, been male), and if land councils see their accountability first and foremost to Aboriginal men, the results are dishearteningly predictable. There may emerge the view, apparent in many claims, that all adult men know more and are therefore better qualified than all women to act as witnesses; even junior men, according to practices developed under this set of assumptions, are treated as if they know more than the most senior women. Aboriginal men may believe that only they are authorized to speak in depth in the context of the hearing. Women are unlikely to have come to understand the specifics of the *Act* and the nature of being a witness. They may not know what their rights as claimants might be, and how they might go about asserting them. They are unlikely to have been proofed to anything like the degree that men are proofed, and anthropologists and lawyers alike may be quite unaware of the depth of knowledge they have to offer. Senior women may not have indicated, or felt that they had the opportunity to indicate, to their legal counsel that they have information which bears crucially on the claim.

One of the most haunting moments of my land rights experience was being taken by the hand by a group of women in a community I was visiting, drawn away to a quiet spot back from the homes, and asked: 'What about that land rights? They going to let women talk for land too, or is it just for men?'

My experience of land claims has been that there is a continuum along which can be situated different women's desire to speak for their country, their desire to demonstrate their status as Law women and their desire to ensure that they as individuals, their group as a whole, and their descendants are understood to be powerful landowners. Individual women, like individual men, position themselves differently, and there are also differences from group to group, region to region. Where women's desires are strong, they have regularly been frustrated.

Women's Evidence

Men's restricted knowledge has been accepted by Land Commissioners as a dimension of Aboriginal culture which they are prepared to respect, and most land claims have included greater or lesser amount of restricted (men only) evidence. Because of the predominance of men in the legal and anthropological positions, this has not appeared to pose a problem. Right from the beginning, however, women's restricted information has been objected to by opposing legal counsel.

Justice Toohey (1982) in his *Report on the Daly River (Malak Malak) Land Claim* discussed his decision about whether or not to receive a submission prepared by anthropologist Diane Bell in conjunction with the women claimants. Dr Bell sought to restrict the submission such that the only man to read it would be the Aboriginal Land Commissioner. Justice Toohey stated:

> It should be clearly understood that if I receive the material it will not necessarily be denied to other parties. As it happens, all counsel participating are male but there are a number of female legal practitioners in Darwin and elsewhere whose services could be enlisted for the purpose of reading the report, just as there are female anthropologists who could be engaged for the same purpose. This may present some practical difficulties. But they are not insuperable. (1982: 86)

His reasoning was by reference to Section 51 of the Act, which reads: 'The Commissioner may do all things necessary or convenient to be done for or in connection with the performance of his functions' (1982: 87).

At one level, the relationship between women's secret/sacred Law and a male Land Commissioner or Judge poses an insoluble contradiction. If women's Law is violated by the presence of men, then a male judge is unlikely to be brought into its presence. Women claimants, unlike men, are thus required to consider an inherent contradiction between the Land Commissioner and the restrictions. Throughout the NT, many women have kept their secret Law secret. This was the decision made by the senior women claimants in the Jasper Gorge Kidman Springs Land Claim (heard in 1988), for example. Like other women in the Victoria River valley, the women in this area have secret/sacred sites, songs, dances, designs and objects; their secret/sacred ritual is owned according to a system of ownership which is coextensive with their system of land ownership; their organization and performance of ritual expresses and authenticates land-owning relationships. These women seriously considered showing the Aboriginal Land Commissioner and all the relevant lawyers and anthropologists involved in the claim a portion of their most secret Law, but at the last moment they decided not to, saying: 'From Dreaming right up to now no man been look that thing. We can't lose that Law' (Rose 1992: 114, 1994).

A strict identification of women's Law with the total exclusion of men, however, overlooks the complex gradations of secrecy in Aboriginal people's skilled and subtle management of knowledge (Rose 1994). The facile contention that if it is not totally secret then it must be totally public has disadvantaged women disgracefully. What matters in land claims, I contend, is not whether women reveal secrets. The important issue is whether women have opportunities fully and freely to give their evidence.

Recent Developments and Potentials

These issues arose in the recent Palm Valley Land Claim and the Tempe Downs Land Claim[3]. Aboriginal Land Commissioner Justice Gray made a series of decisions which radically enhanced the possibilities for women to give their evidence under conditions which facilitate their authority.

The Palm Valley Land Claim was intensely contentious, with three claimant groups, two of whom were in particular and grievous dispute with each other. Unfortunately, no provisions had been made in advance for how evidence might be given so as to spare claimant groups some of the anxiety of having to speak in front of each other, while yet preserving the open hearing which natural justice requires. An extraordinary amount of evidence ended up being given in men-only sessions. Women of disputing claimant groups were in the position that they would never have access to the transcript of evidence given as part of a case against them. The situation for them was intolerable, and they became determined to ensure that they would have an opportunity to speak with the judge themselves. In consultation with a number of these women, and with Diane Smith, the anthropologist employed by the Central Land Council to make some belated efforts at consulting with the women claimants, I suggested a few options which had been trialled in the infamous Wagait dispute.

The women of one claimant group decided after much deliberation to request that they give some evidence to the judge with no men (other than the judge) present. The transcript, they proposed, would be available to be read by the legal advisers and anthropologists involved in the case, but it would not be circulated beyond that set of people. They stated, through their counsel, that these were matters which belong to a restricted domain controlled by women. The women also requested that the NT solicitor, Ms Cullity, not cross-examine them.

Their application not to be cross-examined did not succeed, but the debates around the issue of the exclusion of men other than the Land Commissioner highlight a number of interesting points. The application was put by their counsel, David Avery. Vance Hughston, barrister for the NT Government, objected:

> Mr Hughston: If your Honour is to hear evidence which your Honour is to give any weight to at all, then I would submit in fairness to my clients, that I, as the only experienced land claim counsel amongst my party, should be entitled to hear that evidence… This is not, your Honour, a case of some secret sacred women's matters that cannot be revealed, as I understand it, indeed it is being revealed to your Honour. It is simply a matter that these

> witnesses would feel more comfortable if they could select the group in front of whom they would give their evidence, and if they could select the counsel who can or cannot ask them questions. But in fairness to other participants in an inquiry of this nature unless there are very, very sound religious sacred reasons why it should be done so, in my submissions these proceedings should be as open as possible so that people can have confidence in the conduct of these proceedings, that they are being conducted fairly and openly. If your Honour pleases. (Palm Valley Land Claim Transcript March 1994: 246)

Mr Hughston's introduction of the issue of witnesses choosing to whom they will give their evidence and by whom they will be asked questions did not arise in connection with the men's only evidence, although exactly the same principles would appear to apply. The judge overruled the objection:

> His Honour: The question then really is, is this evidence which involves matters which are secret from men in the ordinary course, in which case it seems to me that I ought to deal with it in exactly the same way as I deal with restricted men's evidence, namely that I should hear it under what I see as a special dispensation, and that I should otherwise exclude men… I feel obligated to hear the evidence in the same way as I would hear restricted men's evidence, and exclude anyone who happened to be female, be they lawyers representing people, anthropologists. My own consulting anthropologist [is] excluded from men's evidence, restricted men's evidence. I feel that I am obliged in the interests of resolving this claim to hear that evidence. (Palm Valley Land Claim Transcript March 1994: 248)

While the application made on behalf of this group of claimant women was to exclude all men other than the Land Commissioner, women of the opposing claimant group did not attend this session. The transcript was, of course, later made available to their lawyer and anthropologist. This other group of women also wanted time with the judge, and a day or two later, they too made an application through their counsel for an equivalent session with the Land Commissioner.

Mr Hughston again objected, this time on slightly different grounds:

> Mr Hughston: If things cannot be revealed to men, they cannot be revealed to men, and once you do reveal them to a man… I cannot see any reason why that cannot be explained to them that it has to be extended to legal representatives of parties, their chosen legal representatives… And again, your Honour, we do not have evidence of what [is] Aboriginal law on this matter, but it just seems an unusual way to approach the matter, to say that it is restricted to women only and then to have a man actually hear it. It just does not seem to me to make any sense, and there is really no evidence which can assist us in working out how it makes any sense. (Palm Valley land Claim Transcript March 1994: 338)

Mr Hughston's appeal to biology—if the information can be imparted to one man, why not to others—is an impressive example of the way in which biological sex has become a category for arguments about restrictions. The biology argument obscures the fact that to be the recipient of knowledge is to be granted a privilege. In the context of land claims, and undoubtedly this will be the case in Native Title cases too, demonstrations of knowledge constitute for Aboriginal people demonstrations of ownership. It must be understood throughout that privately owned knowledge is presented in a public forum because the legislation requires Aboriginal people to demonstrate that they are who they say they are.

Mr Hughston's other point was that the women had not specified Aboriginal Law concerning the restrictions they sought, and that therefore it was impossible for him to know if their application was legitimate in terms of their own Law. The Aboriginal Land Commissioner refrained from engaging in this debate, and thus refrained from positioning himself as an authority on Aboriginal Law. His overruling of the objection represents an important principle for the conduct of hearings:

> His Honour: I have inherited a practice under which restrictions on evidence which is said to be the evidence of men only are freely granted, because, I suppose, commissioners have accepted the word of the representatives of claimants that certain items of evidence are desired to be restricted to men only for good reason, and that without inquiring into the details of

> Aboriginal law [and] the nature of the evidence in a public way, which would be necessary if such an inquiry were undertaken, commissioners have granted the restrictions. I think it would be most unfortunate if evidence from women, which is similarly restricted, were not able to be taken in these claims. Now, the most obvious difficulty about that is that the commissioner must be of one sex or the other, in a biological sense at least, and because I am a man it is easy for me to hear evidence which is restricted to men only. It is obviously not so easy for me to hear evidence which is restricted to women only. In some cases necessity must triumph, and if the women are prepared to make an exception for me to hear that evidence in the interests of the claim and the matters with which the evidence deals, then I am of the view that the first principle is that I should hear it. (Palm Valley Land Claim Transcript March 1994: 339–340)

The Land Commissioner makes a very important point here: Land Commissioners have accepted that Aboriginal Law includes restrictions, and have sought to honour the integrity of the information presented to them without seeking to assert that they exercise authority in relation to Aboriginal Law.

The Tempe Downs Land Claim was heard in November 1994. Again, the claimant women sought restricted sessions during which they could give evidence to the judge. Again, Mr Hughston, acting on behalf of the NT Government, objected. Much the same ground was gone over, and Justice Gray made the percipient point:

> I appreciate that it is short prior notice, and that it is not easy to find representation by experienced female counsel in relation to land claims. But I suppose it might equally be said that if we go on hearing only restricted men's evidence forever, then that is all we will ever do. And it would seem to me to be both unfair to the claimants, and a dereliction of my statutory duty if I were to refuse to hear evidence. (Tempe Downs Land Claim Transcript November 1994: 211)

There the matter might have rested were it not for the fact that the NT barrister expressed his concern that restricted sessions were being used for the presentation of what he thought should be classed as unrestricted information. The exchange between the Land Commissioner and Mr Hughston is instructive in indicating how such concerns can be handled:

> His Honour: I do not have control in advance of what evidence is to be led, and I am invited to make directions restricting evidence. I do that regularly in relation to men, and if I am invited in the same way to do it in relation to women, I propose to do so, provided that they are prepared to make an exception so that I can hear the evidence.
>
> Mr Hughston: Your Honour, could I simply ask, is that irrespective of whether it is of a secret or sacred nature?
>
> His Honour: Well, you would well know, Mr Hughston, that a lot of evidence that is not of a secret and sacred nature comes out in men only sessions… And sometimes there are attempts made to rectify that, to have agreement as to the portion of the transcript of a restricted session… But it is very hard to keep control over the content of the evidence, and I don't feel like stopping people when they are telling me something that could be of importance to them. (Tempe Downs Land Claim Transcript November 1994: 273)

The concerns about whether there might have been information of a non-secret/sacred nature contained in the restricted transcript was dealt with by the provision that the claimants' woman anthropologist would prepare a report, vetted by the claimant women, which would make available to the appropriate men that portion of the information which can properly be communicated to them. To the best of my knowledge there have never been similar provisions giving women access to appropriate portions of evidence given in men-only sessions.

The Tempe Downs Land Claim was a watershed. For the first time extensive evidence was given by women in restricted session with a restricted transcript. For the first time also, a land council (in this case the Central Land Council) made a helicopter available for women's site visits and site evidence. In a world where money talks, this was the first time that money began saying something about gender equity.

In the Tempe claim, a whole group of women was given an opportunity to speak and to show their evidence on their own terms. I was deeply impressed with the authority and strength which the claimant women communicated once they were in control of the context. Those of us who have worked with Aboriginal women and known their strength and authority, only to see that authority overridden in the course of male-dominated land claim processes, will appreciate the momentous possibilities inherent in the reconfiguration of women in land claims.

The implications of Justice Gray's 1994 decision to hear women's evidence in restricted session move in two directions:

1. Within the NT, the land councils and all the other institutions involved in land claims are on notice: women are refusing to be marginalized; if their potential is to be realized, there must be greater investments in research, representation and presentation.
2. The Native Title Tribunal must seriously consider its procedures and practices in order to ensure equality. Based on the NT experience, women who want to have their say as claimants will need to press their rights vigorously in every context available to them.

Acknowledgements The ideas presented here gained form and substance through many conversations with Kimberley Reid, Diane Smith, Diane Austin-Broos and Chantal Jackson. My thanks to each of them for courage, encouragement and support.

Notes

1. There may be historical or cultural reasons why it is appropriate for men to carry the burden of evidence, but this is a matter to be investigated rather than assumed.
2. This aspect of land claims has been commented upon by Peter Sutton (pers. comm) and others; I take it up in Rose (1996).
3. My understanding of these two claims was formed in my capacity as Consulting Anthropologist for the Aboriginal Land Commissioner.

Bibliography

Rose, Deborah Bird. 1992. *Dingo Makes Us Human: Life and Land in an Australian Aboriginal Culture*. Cambridge: Cambridge University Press.

———. 1994. Whose Confidentiality, Whose Intellectual Property? In *Claims to Knowledge, Claims to Country: Native Title Claims and the Role of the Anthropologist*, ed. M. Edmunds, 1–11. Canberra: The Native Titles Unit, Australian Institute of Aboriginal and Torres Strait Islander Studies.

———. 1996. Histories and Rituals: Land Claims in the Territory. In *Mabo, Aborigines and Australia: A Historical Revolution*, ed. B. Attwood. Sydney: Allen & Unwin.

Toohey, John L. 1982. *Daly River (Malak Malak) Land Claim; Report by the Aboriginal Land Commissioner*. Canberra: Australian Government Publishing Service.

Transcript of Proceedings (Auscript). Aboriginal Land Commissioner, Justice P.R.A. Gray. Re: Palm Valley Land Claim, March 1994.

———. Aboriginal Land Commissioner, Justice P.R.A. Gray. Re: The Tempe Downs and Middleton Ponds/Luritja Land Claim, November 1994.

Deborah Bird Rose is a founding figure in the environmental humanities in Australia. She carries out research on multispecies relationships at the edge of extinction. Her most recent book is *Wild Dog Dreaming: Love and Extinction* (University of Virginia Press, 2011). Professor Rose has spent many years learning from Aboriginal people and has written widely on Indigenous kinship with animals. Her books include *Country of the Heart: An Indigenous Australian Homeland* (second edition, 2011, Aboriginal Studies Press) and the prize-winning study *Dingo Makes Us Human* (third edition, Cambridge University Press, 2009). She is an Adjunct Professor in Environmental Humanities at UNSW. Together with Thom van Dooren she founded the journal *Environmental Humanities*. In addition, she serves on numerous advisory boards, including the Association for the Study of Literature, Environment and Culture—Australia-New Zealand and the new journal *The Ecological Citizen*.

6

Ecofeminist Analysis and the Culture of Ecological Denial

Val Plumwood

I will draw on Western philosophy and history to argue that much of the life-threatening crisis that confronts the world in the degradation of the earth's environment can be traced to life-denying elements in the currently dominant culture, the culture of the West. Western culture historically has set the human above and outside the more-than-human sphere, the sphere of nature, which it represents as hyper-separate and lower. This hyper-separation of the human from nature encourages both insensitivity to the damage being done to the earth and the denial of the human species' vulnerability to this ecological damage. Western culture has

---

This chapter was written as a conference paper that Val Plumwood delivered at the 'World Life-Culture Forum: The Transformation of the 21st Century and Life-Culture Sallim' held in Suwon in South Korea from 18 to 21 December 2003, under the auspices of the Kyonggi Cultural Foundation and the World Life-Culture Institute. It is printed here with kind permission from the Val Plumwood archive at the Australian National University and her literary executors, and was not intended by Plumwood to be a final draft. Some of her handwritten notes on the paper have been included in the text. Others, which were deemed by the editors as prompts to her oration, have not been included, but might have suggested further directions in which she wished to develop the work.

V. Plumwood (✉)
University of New South Wales, Sydney, NSW, Australia

L. Stevens et al. (eds.), *Feminist Ecologies*,
https://doi.org/10.1007/978-3-319-64385-4_6

naturalized an ecology-blind conceptual framework of rationality, erected towering illusions of human superiority, and disembodiment—the blind spots of an ancient culture of denial. Human/nature dualism conceives the human as not only superior to but as different in kind from the non-human, which as a lower sphere exists as a mere resource for the higher human one. This ideology has been functional for Western culture in enabling it to exploit nature with less constraint, but it also creates dangerous illusions in denying embeddedness in and dependency on nature.

We see the results of this misunderstanding of human identity in the current reaction of denial of the ecological crisis and refusal to take action to deal with it. Various recent movements and thinkers from both within and outside the Western tradition have challenged this illusory sense of the human as distinct from the sphere of nature, which it can control. These challenges have effected some change in the dominant cultures, but vital cultural transformation hangs in the balance, and global ecological survival may depend on its successful completion.

The Historical Development of Human Apartness

A study of ancient Western philosophy reveals the patterns of thought which treat nature as a disorderly, alien, and inessential sphere in contrast to the humanized, mainly urbanized sphere of reason. Plato's philosophy, for example, treats reason—lodged in a pure realm of immaterial, timeless ideas—as opposed to or threatened by the biological world of nature the corrupted world 'coming to be and passing away'. Ancient Greek thought likewise places the concept of rational 'civilization'—associated with the beginning of urban life—in opposition to the supposedly irrational and chaotic primitive world represented by the primeval forest. Humans, especially male humans, exemplify reason in contrast to nature and animals, of mind in opposition to body. For the Western tradition, these oppositions are gendered. Nature, the body, and the biological 'world of changes' are associated with women and other lower groups such as slaves and animals, in contrast to a strongly separate, higher realm of reason, ideas and 'spirit' associated with elite men. In Plato's philosophy, the earthly world of materiality and embodiment is not only inferior but also corrupting, and those who leave it behind on death pass to a higher and purer realm of immateriality.

These ideas are not confined to ancient philosophy, but were inherited by the dominant Western religious movements of Christianity. In the spirit of the classical Greek tradition of earth denial, Christian ideals of salvation and transcendence subordinated the 'unimportant' earthly world of nature and material life to the next world of heaven, the immaterial celestial world beyond the earth, where non-humans could never go. In the ascent to a better world of spirit beyond earthly, embodied life, matter would ultimately be conquered by the opposing elements of spirit and reason.

With modernity, reason as modern science, began to rival and replace religion as the dominant belief system. Western science replaced but also built on this earlier religious foundation, transforming the idea of conquering nature in death by subordinating nature to the realm of scientific law and technology. Modern science, now with religious status, has tended to inherit and update rather than supersede these oppositional and supremacist ideals of rationality and humanity. In the scientific fantasy of mastery, the new human task becomes that of remoulding nature to conform to the dictates of reason to achieve—on earth rather than in heaven—salvation as freedom from death and bodily limitation. This project of controlling and rationalizing nature has involved both the technological-industrial conquest of nature made possible by reductionist science, and also the geographical conquest of empire which in turn feeds the universal claims of scientific knowledge.

The idea, emphasized in culture, religion, and science, that humans belong to a special sphere apart from nature and animals was of course shockingly challenged by Charles Darwin in his work on the descent of species arguing that humans had evolved from non-human animals. Darwin's insights of continuity and kinship with other animals remain only superficially absorbed in the dominant culture, even by scientists. The traditional scientific project of technological control is justified by continuing to think of humans as a special superior species, set apart and entitled to manipulate the earth for their own benefit. Against the evidence that animals like birds are just as evolved, it is popularly assumed that humans are the apex of creation, more intelligent, more communicative, and much more evolved than other species. The new science of ecology stresses the importance of biosphere services and ecological processes, and the dependence of humans on a healthy biosphere. But the influence of Enlightenment philosophers like Descartes who treat consciousness, rather than embodiment, as the basis of human identity, remains strong and continues in a false consciousness and mode of life which fails to situate human identity, human life, and human places in ecological terms.

Both ecological and ecofeminist analyses then can be seen as indirectly challenging human (ecological) and gender hyper-separation. Many ecofeminists, like other feminists, reject women's traditional place as less than fully human and their consequent inclusion in the separate and inferior sphere of nature opposed to culture. There are two distinct ways to challenge this construction. The first 'earth mother' position accepts the traditional gender separation and the idea that women are part of nature, but reverses the traditional ordering and proclaims that nature is superior to culture. A more thorough challenge, critical ecofeminism (Plumwood 1993), argues that women are no more 'part of nature' than men are—both men and women reside in both nature and culture. This critical position goes on to challenge hyper-separation, both the opposition and polarization of men and women, and that of humanity and nature. Like the first position, it challenges the inferiority of the sphere of nature, but also denies its exclusive link to women. The resulting programme is both feminist and ecofeminist; with feminism, a critical ecofeminism challenges women's exclusion from culture, as the province of elite men who are seen as above the base material sphere of daily life and as entitled to transcend it because of their greater share in reason. Critical ecofeminism challenges the exclusion and distancing of the 'ideal' human, the elite male, from the sphere of nature, ecology, and reproduction to which women have been confined. And like ecology, ecofeminism promotes an ecological consciousness that insists that a truly human life is embedded in both spheres.

The key insight here, as Rachel Carson (1965) understood in the 1960s, and the work of Mary Midgley (1980) and Rosemary Ruether (1975) suggested in the 1970s, is that the hyper-separated conception of humans as beyond animality and 'outside nature' (as a separate and pure sphere which exists 'somewhere else') leads to the failure to understand human vulnerability and dependency on nature that lies behind so many environmental catastrophes, both human and non-human. The environmental problematic is double-sided, because denial of our own embodiment, animality, and ecological vulnerability is the other side of our instrumentalization and devaluation of the natural order. Surviving the environmental crisis thus presents the dominant culture with two linked historic projects of cultural transformation: the task of situating the human in ecological terms and the task of situating the more-than-human in ethical and cultural terms. The first task especially pertains to our contemporary dangerous state of ecological denial.

The Environmental Crisis and the Culture of Ecological Denial

There is no doubt that there is an ecological crisis. It is not a vague future prediction; we are in it right now, and our systems of rationality are not adequate to deal with it—indeed, they have produced it. Let us consider climate change as an example. In the ecological parallel to the Titanic story, we have reached the stage in the narrative where we have received the iceberg warning, and have made the remarkable decision to double the engine speed to full speed ahead and go below to get a good night's rest. A change of course might be bad for business; we might have to slow down, lose time. Nothing, not even the ultimate risk of the death of nature, can be allowed to hold back the triumphant progress of the ship of rational fools.

The often-invoked term 'sustainability' tends to obscure the seriousness of the situation; clearly, no culture which sets in motion massive processes of biospheric degradation which it has normalized, and which it cannot respond to or correct, can hope to survive for very long. We hear of the failure and permanent endangerment of many of the world's oldest and greatest fisheries, the continuing destruction of its tropical forests, the loss of much of its agricultural land and up to half its species within the next 30 years. Seventy-five per cent of the world's fisheries are overexploited. Although the long-term portent of processes potentially disruptive to survival, such as deforestation, global warming, and ocean degradation, is not yet fully grasped, and devastating forms of positive feedback are a real possibility, the attempt to deal with them is being accorded a low priority. This is not a rational course, and if we are told it is, we need to look more carefully at what is meant by 'rational'.

We are mostly going backwards in the key area of containing energy consumption and are facing growing pollution of land, air, and water; the growing problems of the destruction of the forests and the ozone layer; global warming; acid rain and the disposal of toxic wastes; as well as the multiple crises of rationalist agriculture. Our failure to situate ecologically the dominant forms of human society is matched by our failure to situate non-humans ethically, as the plight of non-human species continues to worsen. Rationalized intensive agriculture not only inflicts intolerable living conditions on animals, but increasingly requires massive slaughtering events to stem the disease outbreaks its conditions foster. On the wild side too, primate researchers speak of an 'animal holocaust': we hear of the massive displacement of orangutans, the slaughter of African gorillas, ivory is once again on the world trade menu, and there is a movement to resume the full-scale slaughter of whales. These are the charismatic creatures—for others it is much worse.

For the most part, we know what we have to do to deal with the effects of our actions, but we are not doing it. It is clear that we are dealing with entrenched patterns that will lead to the insupportable degradation of the planetary environment, and that are not open to change by the usual 'rational' processes of demonstration and persuasion. The existing responses of global capitalism to the ecological crisis are not rational. How can an economic system that, say, systematically destroys the earth's protective shield, be considered rational? How can an administrative strategy be rational that leaves these widespread ecological impacts to the self-regulation of corporations that, as Naomi Klein says, cannot even regulate their own book-keeping? (Klein 2002).

The deterioration of the global ecological context of human life demands from our species a clear and adequate response, but we are seemingly immobilized, even though it is clear that at the technological level we already have the means to accomplish the changes needed to live sustainably on and with the earth. So the problem is not primarily about more knowledge or technology; it is about developing an environmental culture that values and fully acknowledges the non-human sphere and our dependence on it, and is able to make good decisions about how we live and impact the non-human world.

We need to understand and explain all of these levels, but I think the phenomenon of *denial*, the failure to understand and confront our own species' vulnerability to ecological damage, is the major barrier to change. The dominant culture fosters certain kinds of delusions of invincibility, which are especially strong among privileged decision-makers. We are confronting not just interest and ignorance here but also various forms of irrationality and illusion that operate at the more general level of dominant culture, resulting in a general insensitivity to our ecological embeddedness. Some aspects of this insensitivity, such as remoteness from ecological consequences, have been greatly worsened under globalization.

Hyper-separation and the Structure of Human/Nature Dualism

It is important then to understand the historical development, logical structure, current expressions of and motivations for the human hyper-separation and human/nature dualism so beloved of the Western tradition. Hyper-separation is an emphatic form of separation that involves much more than just recognizing difference. Hyper-separation means defining the dominant identity emphatically against or in opposition to the subordinated identity, by the exclusion of their real or supposed qualities. The function of hyper-separation is to mark out the Other for separate and inferior treatment. Just as 'macho' identities emphatically deny continuity with women and try to minimize qualities thought of as appropriate for or shared with women, and as colonizers exaggerate differences between themselves and the colonized, so human supremacists treat nature as radically Other. From an anthropocentric standpoint, nature is a hyper-separate lower order lacking any real continuity with the human. This approach stresses heavily those features which make humans different from nature and animals, rather than those they share with them, as constitutive of a truly human identity.

Anthropocentric or human-centred culture often endorses a view of the human as outside of and apart from a plastic, passive, and 'dead' nature, lacking its own agency and meaning. A strong ethical discontinuity is felt at the human species boundary, and an anthropocentric culture will tend to adopt concepts of what makes a good human being, which reinforce this discontinuity by devaluing those qualities of human selves and human cultures it associates with nature and animality. Thus, it associates with nature inferiorized social groups and their characteristic activities; women are historically linked to 'nature' as reproductive bodies, and through their supposedly greater emotionality, indigenous people are seen as a primitive, 'earlier stage' of humanity. At the same time, dominant groups associate themselves with the overcoming or mastery of nature, both internal and external. For all those classed as nature, as Other, identification and sympathy are blocked by these structures of Othering.

This framework of dualisms and hyper-separation is also challenged by anti-colonial thought. Although now largely thought of as the non-human sphere in contrast with the truly or ideally human (identified with reason), the sphere of 'nature' has in the past been taken to include less ideal or more primitive forms of the human, including both women and supposedly 'backward' or 'primitive' peoples taken to exemplify an earlier and more animal stage of human development. Their supposed deficit in rationality or greater closeness to animality invites rational conquest and re-ordering by those taken to best exemplify reason, namely elite white males of European descent and culture. 'Nature' then encompasses the underside of rationalist dualisms which oppose reason to nature, mind to body, emotional female to rational male, human to animal, and so on: progress is the progressive overcoming or control of this 'barbarian' non-human or semi-human sphere by the rational sphere of European culture and 'modernity'. In this sense, a culture of rational colonization in relation to those aspects of the world, whether human or non-human, that are counted as 'nature' is part of the general cultural inheritance of the West, underpinning the specific conceptual ideology of European colonization and the bioformation of the neo-Europes.[1]

An encompassing and underlying dualist and rationalist ideology applying both to humans and non-humans is thus brought into play in the specific processes of European colonization, which has been applied not only to indigenous peoples but also to their land, frequently seen or portrayed in colonial justifications as unused, underused, or empty, an area of rational deficit. The ideology of colonization therefore involves a form of anthropocentrism that underlies and justifies the colonization of non-human nature through the imposition of the colonizers' land forms in just the same way as Eurocentrism underlies and justifies modern forms of European colonization, which understood indigenous cultures as 'primitive', less rational, and closer to children, animals, and nature. The resulting Eurocentric form of anthropocentrism draws on and parallels Eurocentric imperialism in its logical structure; it tends to see the human sphere as beyond or outside the sphere of 'nature', construes ethics as confined to the human (allowing the non-human sphere to be treated instrumentally), treats non-human difference as inferiority, and understands both non-human agency and value in hegemonic terms that background, deny and subordinate it to a hyperbolized human agency (Plumwood 1993).

The colonization of nature through the conception of nature as inferior to the human thus relies on a range of conceptual strategies, which are employed also within the human sphere to support supremacism of nation, gender, and race. The construction of non-humans as 'Others' involves distorted ways of seeing both sameness (continuity or commonality) with the colonized other and their difference or independence. The usual distortions of continuity or sameness construct the ethical field in terms of moral dualism, involving a major boundary or gulf between the One and the Other which cannot be bridged or crossed; for example, that between an elite, morally considerable group and an outgroup defined as 'mere resources' for the first group, which need not or cannot be considered in similar ethical terms. In the West especially, this gulf is established by constructing non-humans as lacking in the department Western rationalist culture has valued above all else and identified with the human—that of mind, rationality, or spirit—or as a lack of what is often seen as the outward expression of mind, viz language and communication. The excluded group—nature, animals, and those humans identified with them—is conceived instead in the reductionist terms established by mind/body or reason/nature dualism, as mere bodies and thus as servants, slaves, tools, or instruments for human needs and projects.

Rationalism and human/nature dualism have helped create ideals of culture and human identity that promote human distance from, control of, and ruthlessness towards the sphere of nature as the Other, while minimizing non-human claims to the earth and to elements of mind, reason, and ethical consideration. Its monological logic of appropriation leads to denials of dependence on the Other in the name of a hyperbolized autonomy, an exaggerated and illusory sense of independence of those human and especially non-human others who support our lives. This denial is functional for appropriation and hence capitalism but leads to relationships that cannot be sustained in real-world contexts of radical dependence on nature.

Humans or Non-humans? Choice and Emphasis

It is important to understand the double-sidedness of human/nature dualism and its impact on both the human and the non-human sides of this radical division. In addition to implicating mind/body, spirit/matter, and related parts of the web of dualisms, ecofeminists have focused on the dualizing and gendering of the human/nature contrast, as variant mappings of reason/nature dualisms in which the (essentially) human is identified with reason, and the more-than-human world is constructed in oppositional and polarized terms as materiality, body, or unreason. In human/nature dualism, the properly human (as reason, coded male) is seen as opposed to and divorced from nature (coded female), as the animal and the ecologically situated body, just as the non-human is hyper-separated from ethics and culture. The failure, characteristic of Western culture, to situate humans (especially elite humans) as ecological and embodied beings is part of the same dualistic construction of the human/nature relationship that also devalues and distances humans radically from the non-human sphere.

The misunderstanding in dominant culture of the human (and consequently of its contrasting category of nature) has been an important ecofeminist theme, for it is precisely by focussing on this point in the web of dualisms that ecofeminist and feminist thinkers have been able to link the categories of environmental and feminist thought to generate a larger narrative. The radical separation that some deep ecologists would seek a remedy for in a personal 'state of being' is much better understood as a major deformation in Western culture that affects the whole understanding of both these key categories, that of the human and that of nature. As such, it cannot be properly dealt with at the level of individual conversion to green uplift aimed at enlarging personal identifications.[2] But we will not 'get' this level of explanation on the importance of human/nature dualism if we confine ourselves to the usual philosophical account, common to both extensionist environmental philosophy and deep ecology of the environmental problematic as the defence of the non-human.

In the mindset of some 'deep' environmental philosophers, ecofeminist struggles concerned with situating human life in ecological terms are decried as 'shallow', while issues of wilderness and the defence of non-humans are treated as 'deep', and are set quite apart from human justice and sustainability issues.[3] This identifies the environmental problematic with what is really only part of it, consideration for other life forms. If the paradigm of environmental activism is wilderness defence, the ethical and ecological failures involved in other kinds of environmental struggles are not addressed, for example, those concerned with nuclear power, herbicides and insecticides, overfishing, desertification, air pollution, unsustainable farming and forestry, unliveable cities, and environmental justice, to name just a few. As I argued in the first section, these are not semi-technical problems of sustainability that can be solved in terms of better political and economic organization. This conventional approach splits the problem into two disconnected parts that have little overlap—lack of compassion for non-humans is identified as the key failure in terms of spirituality and ethics, while human-based ecological problems are treated as failures of political and economic organization. The two halves of the problem are hardly integrated at all, and the environment movement itself appears equally fragmented. What is especially problematic is the way this 'pure' approach goes on to marginalize many highly significant hybrid forms of environmental activism, especially those concerned with environmental justice and with situating human life ecologically.[4]

Ecofeminist analysis reformulates the problem as an outcome or expression of human/nature dualism, a key part of the network of culture/nature, spirit/matter, mind/body and reason/nature dualisms that in Western culture deforms and hyper-separates both sides of what it splits apart. Such a focus on human/nature dualism can give us a fuller, more integrated and coherent conception of the environmental problematic, broadening the narrow focus on non-human and wilderness issues to represent more closely the full range of issues and concerns in real environmental struggles. In this analysis, there are not two unrelated tasks—one of altruism for non-human others and one of a more careful and functional egoism for ourselves—but two closely related tasks, two sides of the same coin of unmaking human/nature dualism by understanding and situating the non-human ethically, and the task of understanding and situating the human ecologically.

The importance of the defence of the non-human world is undeniable, but the environmental problematic is double-sided, with the denial of our own embodiment, animality and inclusion in the natural order being the other side of our distancing from and devaluation of that order. Human hyper-separation from nature establishes a discontinuity based on denying both the human-like aspects of nature and the nature-like aspects of the human, as the denial of the sphere of 'nature' within the human matches the devaluation and denial of nature without. On the other side, the treatment of human concerns as 'shallow' has prevented a more double-sided understanding of anthropocentrism as a problem for humans too, as a factor which prevents us situating ourselves as ecological beings and makes us insensitive to dependencies and interconnections. As usual, it is not only the obvious 'victim', the subordinated party, who is subject to disadvantages and distortions resulting from relationships of domination.

On this analysis, we can cut through the 'prudence versus ethics' debate that has preoccupied so much environmental theory. Our ethical and spiritual failures are closely linked to our perceptual and prudential failures in situating ourselves as ecological beings. To the extent that we separate ourselves radically from nature in order to justify its domination, we lose the ability to respond to it in ethical and communicative terms. We also get a false idea of our own character and location, including an illusory sense of our independence from nature. This is a prudential hazard because it makes us insensitive to ecological limits, dependencies and interconnections.

The historical illusions of human-nature dualism and human self-enclosure are confirmed by contemporary structures at political and economic levels. The current global order follows a logic of the centre, which naturalizes appropriation by privileged groups through a conceptual system in which maximizing egoism and appropriation is rational and the contributions of marginal others, especially ecological others in nature, are rendered invisible, devalued or discounted. Hegemonic definitions of the 'winners' agency and achievement allow denial and backgrounding of the Other's contribution to the outcome, naturalizing appropriation by the hyper-rational 'achiever' as master subject of what the less powerful are or have done, thus justifying and naturalizing the rational achiever's appropriation of their labour and its product.

We arrive then at an interactive explanation in which the ecological denial of contemporary life is overdetermined. It has historical causes and is located in contemporary social and political structures. All contribute to the growing illusion of human life as 'outside nature' separate from and invulnerable to its woes. A feminist framework of analysis shows that the human-centred ideals and conceptions we use to distance ourselves as humans from the non-human world can also explain our failure to understand ourselves as essentially ecologically embodied beings, and shows how they support the dangerous and tenacious illusion of being invulnerable to ecological failure that is perhaps the chief threat to our survival.

Cultural Transformation and Partnership Ethics

As we have seen, Western human/nature dualism is a double-sided affair, destroying the bridge of kinship between the human and the non-human from both ends, as it were. For just as the essentially human is seen as disembodied, disembedded and discontinuous from the rest of nature, so nature and animals are seen as mindless bodies, excluded from the realms of ethics and culture and open to unconstrained exploitation by humans. This double-sided character of human/nature dualism gives rise to two remedial cultural projects which must be integrated. These two projects are the tasks of situating human life in ecological terms and situating non-human life in ethical and cultural terms. Both involve major challenges for the dominant culture. Addressing these tasks and their integration is the aim of a partnership ethics and a partnership re-conception of human/nature relations. A partnership ethics between human and non-human is both possible and necessary. Partnership models consider the needs of both the human and the non-human in a balance of mutual life-giving. I have outlined the philosophical basis of a partnership ethics between humans and nature in my recent book, *Environmental Culture: The Ecological Crisis of Reason* (2002).

The first task of a partnership ethics is to re-envisage ourselves as ecologically embodied beings akin to rather than superior to other animals, and to situate our human lives ecologically, within the sphere of nature. This means giving a high priority to spreading ecological understandings and education. It means abandoning conceptions of rationality like those of the dominant market economy that take no account of our ecological relationships, for situating human life ecologically is the key cultural task of a truly ecologically rational culture. This project of situating human life ecologically involves adapting our lives to the places in which we live and those we live among, evolving ways of life that minimize our ecological impacts, both on ourselves and upon other places and species. It involves keeping track of our ecological impacts, and taking responsibility for the impacts of our lives on other, more remote parts of the ecosystem.

The second partnership task involves enlarging our conception of ethics beyond the human as centre, and the cultural task of recognizing the elements of mind, culture and agency present in animals and the non-human world. Any long-term partnership between two or more agents must be built on reaching some sort of inter-agency, inter-species accommodation and negotiation of mutual needs, the achievement of mutual life-giving between the human and the non-human spheres that replaces the present one-way flow from non-human to human. A partnership project thus presupposes an understanding of the earth, not as exclusive human property to be disposed of for purely human benefit, but as shared with non-human species, elements and forces which are seen as having equal tenure.

Environmental justice is a partnership approach, which has both an inter-generational and an inter-species distributive aspect in terms of the need to share the earth not only with future humans but with other species—including difficult and inconvenient ones. Interspecies distributive justice asks us to provide adequate habitat for species life and reproduction, objecting to the use of so much of the earth for exclusively human purposes that non-humans cannot survive or reproduce their kind. So, it would recognize not just human but also non-human needs as part of the concept of sustainability. Among the skills, ideals and ethical stances, a partnership approach would cultivate are those of sensitivity to, attention to and communication with non-human beings and elements of the world. Partnership ideals of communication aim to replace monologue by dialogue, and exploitation by negotiation.

Many different cultural expressions and elaborations of these projects of partnership with nature are possible, and in many cases it seems that the dominant culture of monologue has much to learn about the possibility of dialogue and negotiation with non-human nature from the wisdom of indigenous and other non-Western cultures.[5]

Notes

1. On bio-information, see Alfred W. Crosby. 1986. *Ecological Imperialism: The Biological Expansion of Europe*. Cambridge: Cambridge University Press.
2. Although these can have a role to play in limited contexts.
3. The mindset that concern with the effect of environmental degradation on humans is 'shallow' and peripheral has been rightly rejected by writers such as Andrew Dobson, who have unfortunately gone on to see this as a reason for rejecting the entire critique of anthropocentrism along with it. See Dobson, Andrew. 1990. *Green Political Thought*. London: Routledge.

4. Another problematic aspect is the insistence of deep ecology on a contextually insensitive prioritizing of non-human versus human issues.
5. The final sentence of this conference paper reads: So I will be much looking forward, during this conference, to learning more about the outlook of my hosts, Korean people and Korean culture.

Bibliography

Carson, Rachel. 1965. *Silent Spring*. London: Penguin.
Klein, Naomi. 2002. *Fences and Windows*. London: Flamingo.
Midgley, Mary. 1980. *Beast and Man: The Roots of Human Nature*. London: Methuen.
Plumwood, Val. 1993. *Feminism and the Mastery of Nature*. London: Routledge.
———. 2002. *Environmental Culture: The Ecological Crisis of Reason*. London: Routledge.
Ruether, Rosemary Radford. 1975. *New Woman New Earth*. Minneapolis, MN: Seabury.

Val Plumwood (b.1939–d. 2008) was a trailblazing ecofeminist, environmental philosopher and activist who was named one of the *Fifty Key Thinkers on the Environment* (Routledge 2001) alongside Spinoza, Thoreau and Rousseau. Her book *Feminism and the Mastery of Nature* (1993) is considered a classic of ecofeminist and environmental theory. Her co-authored book with Richard Routley, *Fight for the Forests* (1973), was a landmark exposé of the practices of the Australian forestry industry in the 1970s and became instrumental in changing the Australian government's forest policy. In the 1970s, Plumwood and Routley bought and protected a forest property, Plumwood Mountain, above the Clyde River Valley in Southern New South Wales, Australia. Since Plumwood's death in 2008, Plumwood Mountain continues to be protected by a conservation trust.

Part II

Ecofeminist Currents

7

From *The Female Eunuch* to *White Beech*: Germaine Greer and Ecological Feminism

Lara Stevens

Although she is still publicly active in the twenty-first century, iconoclastic feminist Germaine Greer is best known for *The Female Eunuch* (1970), her radical critique of patriarchy. Greer has recently published *White Beech: The Rainforest Years* (2013), a memoir that recounts her experience of ageing as she contends with the challenges of her ambitious, long-term project of protecting and rejuvenating 60 hectares of Australian land from 'steep rocky country most of it impenetrable scrub' (1) to its original rainforest ecology. We might think of these books as potential bookends to Greer's career spanning almost half a century: *The Female Eunuch*, her revolutionary call to arms, and *White Beech*, her reflective and spiritual encounter with her piece of Australian rainforest in south-east Queensland. The former was written in a time before gender or women's studies existed as a discipline in universities; the latter amidst an upswing of university departments devoting teaching and research resources to the Environmental

L. Stevens (✉)
The University of Melbourne, Melbourne, VIC, Australia

L. Stevens et al. (eds.), *Feminist Ecologies*,
https://doi.org/10.1007/978-3-319-64385-4_7

Humanities. Yet these books point to more than just Greer's capacity to anticipate areas of urgent public interest. Both works share Greer's sardonic and superior tone, her activist provocations, her challenges to the status quo, as well as her insatiable curiosity as she seeks out knowledge in areas as diverse as genetics, botany, Indigenous cultural practices, and English etymology.

The Female Eunuch was a polemic that aimed to incite a revolution in women's sexual lives at the historical junction of protest and revolt in the West in the late 1960s. It coincided with the inception of second-wave feminism and with two other groundbreaking publications in America in the same year: Shulamith Firestone's *The Dialectic of Sex* (1970) and Kate Millett's *Sexual Politics* (1970). All three texts were written by women with academic training, but, of the three, *The Female Eunuch* was the most accessible to a non-academic audience because it employed a journalistic style and language, first person anecdotes, and less philosophical and literary references. It became far more widely known than other contemporaneous feminist texts because Greer's audacious public statements and striking physique intrigued and excited the malestream press. As feminist cultural historian Mary Spongberg writes:

> Unlike contemporary American feminists such as Kate Millett and Robin Morgan, who deliberately avoided the press, one of Germaine Greer's primary aims was to appeal to it. In her view, one of the great failures of the women's movement was its refusal even to attempt to harness the power of the media. (1993: 408)

Spongberg explains that Greer's pro-male stance and frequent derisive comments about other feminists 'enabled the press to engage in blatantly misogynistic ranting under the guise of championing feminism' (1993: 412).

The Female Eunuch has, to date, received surprisingly little scholarly attention despite being one of the most widely read feminist texts of the twentieth century (Taylor et al. 2016: 1). As Sybil Nolan has noted, *The Female Eunuch* has never been treated as a serious scholarly work in the academy—there is no 'Greerology'—yet Greer is a well-known public figure in Australia and the United Kingdom (1999). The project of rectifying this gap began in a recent edition of the journal *Australian Feminist Studies* (Vol. 31, Issue 87, 2016) themed around Greer, edited by Anthea Taylor, Maryanne Dever, and Lisa Adkins. This edition was compiled in response to the 2015 opening of the Germaine Greer Collection, her extensive archive housed at the University of Melbourne.[1]

Greer has always been a controversial figure within feminism and she has never identified easily with her sisters (Spongberg 1993: 414). Likewise, her ideas, public commentary, and publications have never fitted straightforwardly into feminism's so-called waves. Although in the opening pages of *The Female Eunuch* Greer positions the book as 'part of the second feminist wave' (1970: 11), historian Marilyn Lake notes Greer's openly hostile and antagonist attitude to other feminist activists of the time. In rough drafts and revisions of the book proposal for *The Female Eunuch* (housed in the Germaine Greer Collection), for example, Greer describes the women's liberation movement of the 1960s as 'tiny, privileged and overrated' (Greer in Lake 2016: 8). For Greer, the women's movement was not doing enough to advance the cause of the every-woman and was not radical enough to snap women out of their passivity or liberate their repressed sexuality. She was particularly critical of Betty Friedan's *The Feminine Mystique* (1963) for its narrow appeal to white, educated, middle-class American women (Lake 2016: 12). As Lake notes, Greer was determined to 'draw fire from all the articulate sections of the community' (12), not simply from the privileged few. Greer's desire to appeal to a wide audience, and not only the educated elite, might explain the voluminous positive and negative responses to *The Female Eunuch* both within and beyond the Western world (see, e.g. Spongberg 1993: 414 and Greer 2010: np).

Despite Greer's proclaimed disapproval of various feminists and feminist movements, Lake notes that Greer's thought has a lot in common with two of the most important feminist thinkers to precede her. Lake writes:

> For all Greer's dismissal of earlier feminist writers and her bluestocking forebears in the draft proposals, her arguments actually bore a striking similarity to those enunciated by Mary Wollstonecraft in the late eighteenth-century and Simone de Beauvoir in the mid-twentieth century ... For all its declared sensationalism, *The Female Eunuch* was in many ways a deeply conventional feminist text. (2016: 17)

For Lake, Greer's refusal to identify strongly with other feminist movements or thinkers was somewhat disingenuous given her debt to de Beauvoir and Wollstonecraft. Yet, Lake equally acknowledges the unique impact that Greer's landmark book had on future generations of women and feminist thinkers (1999: 228). To better understand *The Female Eunuch*, its structure, its critique, as well as its enduring legacy, I apply ecofeminist thought on the basis that such a reading reveals aspects of the work that were ahead of their time, advancing, rather than simply rehearsing, the critiques of patriarchy made by key feminist foremothers. This is not to say that *The Female Eunuch* is a work of ecofeminism, but, rather, that aspects of its critique can be better appreciated using ecofeminist thought and that this mode of analysis carries through to Greer's writing on ecology and natural history in *White Beech*.

In this chapter I show how the theories of preeminent ecofeminist philosophers Val Plumwood, Ariel Salleh, Deborah Bird Rose, Kate Rigby and Freya Mathews offer useful ways to understand both the form and content of Greer's analysis of patriarchy. I employ Plumwood's ecocritical viewpoint that conceives of 'oppression as a network of multiple interlocking forms of domination linked by a flexible, common ideology and structure of identity' (1994: 78) to argue that *The Female Eunuch* advocates the radical overturning of these complex networks of domination. In the 1990s, Plumwood names similar modes of critique 'critical ecological feminism' or 'environmental feminism' (1993: 1). Further to this, I show how such critical ecological feminism resurfaces in *White Beech* and is integral to its account of the history of how and why Greer's rainforest property became degraded, overrun with introduced flora and fauna, and stripped of its original custodians.

The Female Eunuch: Dismantling the 'Conceptual Apparatus'

The Female Eunuch can be usefully read within the framework of critical ecological feminism because it is interested in linking the power structures of patriarchy, colonialism, feminism, and capitalism to cultural and scientific representations of Indigenous Australians and the land. Although the writing and publication of *The Female Eunuch* precedes the use of the term 'ecofeminism' (which is thought to have originated in the activism and writings of Françoise d'Eaubonne, particularly her book *Le féminisme ou la mort* (1974: 213–252)), it is not only the content of *The Female Eunuch* that suggests what becomes known as ecofeminist concerns. Its mode of argumentation relies on demonstrating the intertwining of systems of oppression and domination, a key characteristic of ecofeminist methodology. Such a reading shifts how we might appreciate *The Female Eunuch* through discourses that were not available to Greer or her readers at the time the book was written and published.

The Female Eunuch begins with an attack on the 'conceptual apparatus' that falsely presents the sexes as a 'polarity and a dichotomy in nature' (Greer 1970: 25). This, she argues, perpetuates myths that emphasize the differences between the sexes rather than their similarities. In the 'Gender' chapter Greer writes:

> the animal and vegetable worlds are not universally divided into two sexes, or even into two sexes with the possibility of freaks and indeterminate types; some lucky creatures are male and female by turns some fungi and protozoa have more than two sexes and more than one way of coupling them. (25)

Here she draws upon the complexity of the natural world to highlight the shortcomings of popular and scientific characterizations of gender in the human species as neatly and unproblematically differentiated. She then compares animals and microorganisms that change sex over time, or are hermaphrodites, with the rigid male/female binary system applied to humans to show how our species' sex difference is frequently 'stressed and exaggerated' (26).

Greer goes on to note that the epistemological frameworks that divide the human sexes so determinately have historically provided unfounded scientific justifications for the practice of eugenics. She writes:

> Nazi anthropologists maintained that the secondary sexual characteristics are more highly developed in more highly evolved species, pointing out that Negroid and Asiatic types frequently had less defined secondary characteristics than Aryans. (Greer 1970: 25)

Here she cites H. H. Ploss and M. and P. Bartels (translated into English by E. J. Dingwall in 1935) whom she describes as the 'embodiment of anthropological and ethnological prejudice' (332n). The 'conceptual apparatus' that Greer attacks in this chapter is Enlightenment scientific rationalism taken to its extreme. She shows how sex binaries can be exploited to inscribe and reinforce the dominant gender ideology in ways that are detrimental, not only to women, but also to particular racial and non-heterosexual minority groups. The historical over-emphasis and misrepresentation of these connections between nature and other 'others' by scientists, anthropologists and philosophers are what ecofeminists such as Plumwood (1993), Haraway (1989), Salleh (1984a) and others vigorously demonstrated in the 1980s and 1990s.

Ecofeminists have long shown how a feminist revolution is unthinkable without a complete overturning of deeply embedded epistemologies that differentiate human-nature relations as antagonistic dualisms and hierarchies that relegate both nature and the feminine to a space beneath the realm of reason, culture, and patriarchy (Plumwood 1993, 1994); (Warren 1994); (Haraway 1989). Freya Mathews frames the dualisms as such:

> dualism constitutes a full blown ideology which interprets the world in terms of dichotomous pairs of qualities, such as active/passive, light/dark,

> mind/body, reason/emotion, and Culture/Nature. ... The entire system exists for the purpose of legitimating the inferiorization of the feminine and all things traditionally associated with it. (1994: 165)

Ecofeminists see the patriarchal domination of nature and women as arising from the same ideological foundation (Ruether 1975; Plumwood 1993, 1994; Warren 1987, 1994), dating back to the Ancient Greek philosophers, but exacerbated during the scientific revolution of the Enlightenment (Plumwood 1993). Greer's critique of this binary system as an oppressive 'conceptual apparatus' and her recognition of nature as polymorphous radically precede later ecofeminist depictions of human-non-human relations as complexly networked and interrelated.

Ecological feminism as a field is not only concerned with the ways in which the environment impacts women's lives, it also traces historically and philosophically the points at which attitudes towards women and attitudes towards the natural world converge within a broader set of systemic oppressions. Given the Western Enlightenment project of ordering 'chaotic' nature, Plumwood argues that the oppression of women and the destruction of nature are linked by an ideology of control that privileges man as the embodiment of reason over nature (1994: 64, 74). For Plumwood, unravelling the systems and attitudes that keep women and nature subordinate can only be achieved by acknowledging their shared history of oppression.

The Female Eunuch conceives of women's oppression as a network or system of multiple and interlocking dominations. The chapters cover criticism of the detrimental effects of capitalism on women, representations of women in popular romance novels, male scientific discourses on female body parts, the role of technology, war and violence in society, women in the labour force, parental attitudes towards female children, the power of the media, and Freudian theories of female psychology, to name a few topics in Greer's sight. The ways in which these discussions overlap and reoccur within the broader analysis of patriarchy suggest that Greer sees these problems as interrelated. As such she paves the way for the relational thinking of many ecofeminist scholars, such as Kate Rigby, who notes:

> what distinguishes ecofeminist thinking from an environmentally concerned feminism, or, conversely, a pro-feminist environmentalism, is the claim that there is some kind of inherent or structural connection between the patriarchal domination of women … and the ecologically destructive exploitation of the earth. (1998: 144)

Greer's discussion of human attitudes towards the natural world and the role of biological science in characterizing species in particular ways reveal how attitudes towards nature and women are historically and ideologically linked. Through this discussion Greer pre-empts Plumwood's (and other ecofeminists') questioning of binary thinking and her desire for reason and science to be reconfigured 'in less oppositional and hierarchical ways' (Plumwood 1993: 4).

The 'Smorgasbord of Ideas'

In *The Female Eunuch*, Greer's rhetorical mode of argumentation is impassioned, performative, iconoclastic, and often didactic. Her prose moves between first person anecdotes and historical accounts that aim to politicize readers and raise their consciousness about pertinent feminist issues. Her mode of writing employs the rhetoric of a public intellectual and activist. For example, she writes that: 'Revolution is the festival of the oppressed' (1970: 330) and concludes the book with the rhetorical provocation to her readers: 'what *will* you do?' (331). Ian Britain describes *The Female Eunuch* as 'a book of outrage: an exposé, a jeremiad, a manifesto' (1997: 139). The polemical aspects of *The Female Eunuch*, combined with the fact that Greer wrote the work in the immediate wake of the May 1968 revolutionary movements in Europe, suggest that it shares something in common with twentieth-century manifestos and their earlier prototypes, such as Karl Marx and Friedrich Engels' *The Communist Manifesto* (1848) or Filippo Tommaso Marinetti's *Futurist Manifesto* (1909). The use of narrative in *The Female Eunuch*, in particular its exploration of women's oppression under the categories of 'Body', 'Soul', 'Love', 'Hate', and, finally, 'Revolution', owes a debt to the rhetorical form of the Marx and Engels template. It is uncompromising in its call for total liberation stating: 'reforms are retrogressive. The old process must be broken, not made new' (Greer 1970: 331). Greer is not a liberal or equality feminist. Rather, like so many ecofeminists, her brand of feminism demands a dismantling of the entire system of patriarchy and its links to capitalism and scientific rationalism.

Salleh notes that ecofeminism's structural vantage point gives it a unique capacity to connect divergent areas of injustice. She writes that:

> No other perspective—liberalism, socialism, feminism, environmentalism—can integrate what ecofeminism does: why the Roma people are still treated like animals; why women do 65 per cent of the world's work for 10 per cent of its wages; [...] why the Earth itself is manipulated as a weapon of war. (Salleh 2014: xii)

Commentators on *The Female Eunuch* often draw attention to the variety of subject matter with which the work engages. Ann McGrath describes *The Female Eunuch* as a 'smorgasbord of ideas and information', full of 'delectable morsels' that make it highly readable and strategically marketed (1999: 183). McGrath notes that the tone of Greer's writing is at once polemical, didactic, preaching, hectoring, banner-waving, and often superficial in its treatment of each topic but never lacking passion (184). McGrath's description of *The Female Eunuch* as a 'smorgasbord of ideas and information' reflects the book's mode of analysis. Greer's reticence in aligning herself with activist movements in the 1960s and 1970s and her often individual and controversial positions on a range of feminist issues today show her scepticism towards liberalism, socialism, feminism, environmentalism, indeed any 'ism' that constrains her to a singular set of beliefs. Her unique brand of feminism, that found the 1960s' women's liberation movement to be 'tiny, privileged and overrated' and continues to refuse to neatly identify with any particular feminist or postfeminist labels, shares the ecofeminist refusal to be boxed into a position. This opens out her critique of patriarchy in *The Female Eunuch* (and her feminist commentaries since) to a broader structural critique but makes her position idiosyncratic such that her ideas have been known to alienate and frustrate other feminists.[2]

In a more critical response to *The Female Eunuch*, writer and journalist Laura Miller describes it as a:

> fitful, passionate, scattered text, not cohesive enough to qualify as a manifesto. It's all over the place, impulsive, and fatally naïve—which is to say it is the quintessential product of its time. (1999: np)

Miller's critique overlooks the strategic nature of the book's form and structure for purposes other than tactical marketing. *The Female Eunuch*'s 'impulsive' connections appear 'all over the place' to Miller, yet, I suggest that they prefigure Greer's understanding of women's oppression and repression as rooted in structural inequalities. As I have shown, this kind of thinking is later identified as a critical ecological feminist methodology. Greer deploys such an approach in *The Female Eunuch*, decades before environmental ecology was thematically at the forefront of her writing and her activism as it is in *White Beech*.

White Beech: Secret Women's Business

White Beech describes the beginnings of Greer's long-term project of rehabilitating a piece of land that she purchased in the Numinbah Valley in south-east Queensland in 2001. She refers to it as the Cave Creek Rainforest Rehabilitation Scheme or CCRRS. Although Greer is best known for her writings on women, Shakespeare, seventeenth-century literature and literary women, her interest and involvement in environmental activism, natural history, gardening, and the botanical sciences date back at least to the early 1980s. It was at this time that she moved from London to a three-acre property in rural north-west Essex where she planted a woodland area. She has long been actively involved in the British organization Buglife, the only European NGO devoted to the conservation of Britain's rarest invertebrates including bees, beetles, worms, woodlice, jumping spiders, and jellyfish, and was made its Vice President in 2004 and President in 2006 (Buglife 2017). *White Beech* recounts her search for a piece of land in Australia that she could regenerate and protect. Above all, it tells the story of finding Cave Creek and her attempts to uncover not only its natural history but also its ancient human history. Greer's anti-binary or ecofeminist thinking reappears in *White Beech* through her analysis of the history of environmental degradation at Cave Creek and the current and future challenges for restoring the land's biodiversity.

Greer explains that as soon as she purchased the land she went in search of its traditional owners. She recounts how she was met, however, with either silence or confusing, contradictory accounts of Indigenous peoples' presence on the land. After a complicated search that led to many dead ends, Greer developed a hypothesis about a particular part of the rainforest called 'Natural Bridge', a famous rock and popular tourist destination that forms a cave, penetrated by a powerful waterfall. She writes:

> My guess is that the cave was a place of serious women's business, even of pilgrimage in time of special need. Infertility. Unwanted pregnancy. Maybe even infanticide. (Greer 2013: 128)

Greer uses historical accounts of Natural Bridge and her own reading of its landscape to argue that the secrecy of the local Indigenous people around the original name of the site and its role in Aboriginal culture is because it is a sacred or cursed place. She arrives at this conclusion by reading Natural Bridge's unique topography: its deep gorge, arching rock, low cavern, and the thunderous noise of the waterfall. Greer describes the combination of these elements as 'an image of titanic intercourse' (127), meaning perhaps that the place gives the impression of a life-creating encounter. She believes that such an example of the sublime power of nature means that the site must have had or continues to have great cultural significance for the local Indigenous populations.

Greer's reading of the topography of Natural Bridge is informed by her interest in Indigenous land rights claims. In particular, she mentions claims based on 'secret women's business', a term that she describes as 'the most ridiculed three words in a nation given to ridiculing anything it cannot understand' (Greer 2013: 127). Greer reminds readers of the case of the Ngarrindjeri women in the 1990s, whose land rights claim to a sacred burial site on Hindmarsh Island was initially dismissed as a fabrication by the Royal Commission (1995) headed by former Supreme Court judge Iris Stevens, but was later upheld. For Greer, the 'whitefellas' discredited the claim of the Ngarrindjeri women largely because of the plaintiffs' status as both Indigenous and women. When the workmen who were redeveloping the wharf dug up the skeletal remains of a Ngarrindjeri woman and her daughter, the law and its patriarchs finally had the scientific proof they desired but it came at great expense for the women who held that land to be sacred.

The complex connections between Indigenous spirituality, ceremonies, migration routes, the topography of the land, animal habitats, and natural occurrences such as bush fires are poorly understood by white Australians. For Greer, the white Australians' inability to read the land makes them blind to its spiritual and sacred significance. She writes of the Hindmarsh Island claim that: '[t]he secret was manifest in the topography if only the developers had had eyes to see' (Greer 2013: 127). Anthropologist Deborah Bird Rose has long stressed the importance of recognizing Indigenous women's claims to land through their connection to sacred ritualized practices:

> women … have secret/sacred sites, songs, dances, designs and objects; their secret/sacred ritual is owned according to a system of ownership which is coextensive with their system of land ownership. (1995: 4)

Greer's attempt to understand the Australian landscape and its 'narratives' or 'voices' outside of the myopia of Western cultural and scientific frameworks is linked to her concern that the rights of Indigenous women be taken seriously in the Western legal system and Australian culture more broadly. In *White Beech*, Greer's concern for Indigenous women and their relationship to the land is part of a broader critique of white settler treatment of Indigenous people.

Plumwood's criticism of Western scientific and philosophical systems is interested in showing how patriarchal 'mastery' over nature goes hand in hand with controlling women and Indigenous peoples (1994: 64). She explains that, dating back to the explorations of Christopher Columbus, colonizers emphasized Indigenous people's association with 'nature' and their 'irrationality'. By describing Indigenous people as 'uncivilized', colonizers justified invasion, enslavement, and slaughter by 'rational' and 'civilized' Westerners (Plumwood 1994: 75). In the case of Australia, the white settlers' claim of '*terra nullius*' was used as a means to justify the invasion and seizure of land from Indigenous communities. Rose's (1995) work shows that, despite the gains of the Native Title Act (1993), the gender biases in the Australian legal system mean that Indigenous women continue to feel the effects of the powerful myth of *terra nullius* most acutely. In an attempt to resist a long history of patriarchal and capitalist exploitation of the land and the instrumentalization of Indigenous people and the natural world, Greer insists throughout *White Beech* that she does not own Cave Creek. She writes: '[i]t would never have occurred to me that my whitefella freehold title endowed me with proprietorial rights' (Greer 2013: 114). She renounces ownership not only by inviting Indigenous people to live on the land, but also by signing the property rights over to the Friends of Gondwana Rainforest, a registered charity committed to preserving rainforests in the ancient continent of Gondwana and which will ensure the continuation of the rehabilitation of Cave Creek after her death.

In *White Beech*, Greer gives an account of how white settlers reinforced the idea of *terra nullius* on a symbolic level by renaming the natural world. She pays particular attention to the ways in which the natural sciences used naming and species classification as a mode of control. The internationalism of taxonomy and scientific naming was enabled by the broader project of European colonization, linking environmental and colonial oppression (Keller 1985: 17). In *White Beech*, Greer shows that the European botanists' naming of the flora, fauna and natural sites at Cave Creek reflected racist and sexist attitudes—white settlers ignored Indigenous names for the land, its seasons, and its creatures by imprinting white colonizer names on everything they found. Furthermore, they used patronymics as a means by which to assert patriarchal authority and erase or discredit existing knowledge. Allan Cunningham, for example, named many of the trees that populate Cave Creek after himself including the Hoop Pine, *Araucaria cunninghamii*; the Bangalow Palm, *Archontophoenix cunninghamiana*; the *Casuarina, Allocasuarina cunninghamiana*, the Native Tamarind, *Diploglottis cunninghamii*; and the Brown Beech, *Pennantia cunninghamii* (Greer 2013: 222). Greer argues this was a desperate attempt by European botanists to assert control over the harsh and untamable Australian climate and landscape that did not respond well to European species and agricultural methods.

Greer comments that, historically, the use of patronymics has been particularly prevalent in the discipline of botany. She scornfully lists the many inaccurate names that male botanists imposed upon Australian species (2013: 210–237). She points out the contradiction that although patronymics were widely used, botany was one of the few professions in which it was socially acceptable for women to participate. She writes:

> Many women botanized, and bred and grafted horticultural varieties, but the intellectual conquest and ordering of the vegetable world cannot have held the same appeal for them as it did for men. The number of women who authored plant names is pathetically small; not only did very few women do it, they only did it once or twice, whereas men like Hooker, Bentham and Mueller authored literally hundreds of names. (235)

Greer laments that the significant contributions of female botanists as well as the women who lived on and worked the Australian land are often omitted from history by the dominance of patronymics. She shows that the use of patronymics in Australian botany is a mode of mastery and ownership in the history of male, colonial, hubristic attitudes towards the environment. Greer's critique of patronymics in Australian botany suggests that the language and discursive practices of science reinforce the gender and racial biases of society more broadly and affect how we think about and relate to that matter.

Greer's Ecological Self at Cave Creek

Greer's work with the rainforest over the last decade has not only had an effect on the land but has altered her view of herself. Where scientific and colonizing man saw himself at the centre of the world—stamping his name in history and his will on what he perceived to be a hostile land—Greer claims that her close experience with the flora, fauna and land at Cave Creek led to an obliteration of her ego. Summarizing her journey of learning about the forest she writes:

> My horizons flew away, my notion of time expanded and deepened, and my self disappeared. (Greer 2013: 2)

Greer's experience suggests a relationship to selfhood that is more akin to Freya Mathews' concept of an 'ecological self' (1991). Drawing from some of the ideas of 'deep ecology' that value the non-human world independent of its usefulness for human purposes, Mathews' theory sees human and non-human life forms as each having 'intrinsic value' that does not privilege one over the other (1991: 120). Instead, she views systems in 'feedback loops' where the part conditions the whole and the whole conditions the part (143). She writes:

> Holistic nesting of a self in a wider self-system means a relative identification with that system. Because the self stands in relations of ecological interdependence (direct or indirect) with the element of that wider self, those elements (or its relations to them) are logically involved in its identity. (144)

Greer employs similar ecological metaphors to Mathews. Whereas Mathews sees selfhood as a 'nesting' within a broader system, Greer's 'horizons' drift apart through her connection to the land. Greer claims that her former sense of self, based on a proudly forthright and active sexualized ego, is replaced by an identity in which she is 'the sister of its [the forest's] mosses and fungi, its mites and worms' (Greer 2013: 2). Greer adopts an ontological view that puts her on equal footing with microorganisms—an ecological self within a broader interdependent system.

Greer sees biodiversity as key to what she describes as an imaginative force in a codetermining environment that is constantly shaping and reshaping the evolution of millions of macro- and microspecies (2013: 3). Such a view of nature means contesting species bias. She writes:

> It makes no sense to start trying to save a disappearing plant without dealing with the conditions that are causing its disappearance, and that requires restoration of the plant community of which the rare plant is a member. (107)

In this observation, Greer sees the plant within a web of interconnections, a broader 'plant community', which is to say, all the other matter that makes life possible right down to the water, minerals and microbes in the soil. She considers her work protecting and rejuvenating endangered native species at Cave Creek not as atomized projects but, rather, as a reestablishment of the land's biodiversity. Her concern that we learn to 'respect the entire system, not just the bits we consider cute or useful' (301), aligns with contemporary ecofeminist and ecosophical challenges to traditional ways of thinking about human-non-human identity, subjectivity, and agency.

Conclusion

The Female Eunuch is a pioneering and far-sighted feminist text that can be better understood through the lens of ecofeminist scholarship that has emerged in the decades since its publication. It employs formal structures and modes of argumentation that characterize the power structures of patriarchy, colonialism, and capitalism as webbed and interrelational. Salleh notes of ecofeminism in the 1980s: '[t]he accidental convergence of feminism and ecology at this point in time is no accident' (1984b: 344). Similarly, the 1960s and 1970s saw the rise of environmental movements, often intertwined with anti-war, anti-nuclear energy, and anti-weaponry activism; anti-uranium mining (which was being linked to miscarriage, birth defects, and leukaemia); civil rights movements in America; and gay and lesbian rights activism alongside the feminist movement. It was this same vibrant and politically charged historical moment in Europe, America, and Australia in which the ecofeminist movement too was gathering momentum. Looking back at *The Female Eunuch* as a twenty-first-century feminist, it appears to be no accident that Greer's thematic concerns and her analysis of interleaving power structures and their oppressive effects on the human and non-human world presage many of the concerns of ecofeminist theorists over the last four and a half decades.

In *White Beech*, Greer links patriarchy, colonialism, and capitalism to her politicized reading of the rainforest topography of Cave Creek as a site of secret women's business, the history of scientific naming of botanical species using patronymics, and the difference in attitudes between white and Indigenous Australians towards land ownership. Despite different emphases in *The Female Eunuch* and *White Beech*, these two texts share a common interest in exposing systemic oppression. Neither text characterizes its critical focus—women's oppression or environmental damage—as singular, atomized problems within society. Both texts employ formal structures and modes of argumentation that illuminate how power structures are interconnected in ways that pre-empt and may even play a part in motivating the work of critical ecological feminism.

Notes

1. The Germaine Greer Collection includes documents predominantly from 1957 to 2010 that chart the life and career of Germaine Greer. The kinds of documents in the collection include: drafts of her books and articles, research files, photographs, sketches, books, correspondence, fan mail, magazines, digital audiovisual recordings, awards, video and text media (The Germaine Greer Collection online).
2. For example, her comments on transgender people (John 2016) and on former Australian Prime Minister Julia Gillard are among her more recent controversial remarks in the media and public debate (Radio National 2017).

Bibliography

Britain, Ian. 1997. *Once an Australian: Journeys with Barry Humphries, Clive James, Germaine Greer and Robert Hughes*. Melbourne: Oxford University Press.

Buglife: The Invertebrate Conservation Trust website. 2017. Accessed March 28, 2017. https://www.buglife.org.uk/

D'Eaubonne, Françoise. 1974. *Le Féminisme ou la mort*. Paris: Pierre Horay.

Firestone, Shulamith. 1970. *The Dialectic of Sex: The Case for Feminist Revolution*. New York: Morrow.

Friedan, Betty. 1963. *The Feminine Mystique*. Middlesex, England and Victoria, Australia: Penguin Books.

Germaine Greer Collection, The. 2017. University of Melbourne. Website. Accessed March 15, 2017. http://archives.unimelb.edu.au/germainegreer/home

Greer, Germaine. 1970. *The Female Eunuch*. St Albans, Herts: Paladin.

———. 2010. Better Half Still Battling. *The Australian—The Australian Literature Review*. Online, May 5. Accessed March 28, 2017. http://www.theaustralian.com.au/arts/books/better-half-still-battling/story-e6frg8nf-1225860734719

———. 2013. *White Beech: The Rainforest Years*. London: Bloomsbury.

Haraway, Donna. 1989. *Primate Visions*. New York and London: Routledge.

John, Tara. 2016. Germaine Greer Defends Her Controversial Views on Transgender Women. *Time*. Online, April 12. Accessed March 28, 2017. http://time.com/4290409/germaine-greer-transgender-women/

Keller, Evelyn Fox. 1985. *Reflections on Gender and Science*. New Haven and London: Yale University Press.

Lake, Marilyn. 1999. *Getting Equal: The History of Australian Feminism*. St Leonards: Allen & Unwin.

———. 2016. 'Revolution for the Hell of It': The Transatlantic Genesis and Serial Provocations of *The Female Eunuch*. *Australian Feminist Studies* 31 (87): 7–21.

Marinetti, Filippo Tommaso. 1973 [1909]. The Founding and Manifesto of Futurism. In *Documents of 20th Century Art: Futurist Manifestos*, ed. U. Apollonio and trans. R. Brain, R.W. Flint, J.C. Higgitt, and C. Tisdall, 19–24. New York: Viking Press.

Marx, Karl, and Friedrich Engels. 1952 [1848]. *Manifesto of the Communist Party*. Moscow: Progress Publishers.

Mathews, Freya. 1991. *The Ecological Self*. London and New York: Routledge.

———. 1994. Relating to Nature: Deep Ecology or Ecofeminism? *The Trumpeter: Journal of Ecosophy* 11 (4): 159–166.

McGrath, Ann. 1999. *The Female Eunuch* in the Suburbs: Reflections on Adolescence, Autobiography and History-writing. *Journal of Popular Culture* 33: 177–190.

Miller, Laura. 1999. Germaine Greer. *Salon*. Online, June 23. Accessed May 3, 2005. http://www.salon.com/1999/06/22/greer/

Millett, Kate. 1970. *Sexual Politics*. New York: Doubleday.

Nolan, Sybil. 1999. Tabloid Women. *Meanjin* 2: 165–177.

Plumwood, Val. 1993. *Feminism and the Mastery of Nature*. London and New York: Routledge.

———. 1994. The Ecopolitics Debate and the Politics of Nature. In *Ecological Feminism*, ed. K.J. Warren. London and New York: Routledge.

Radio National. 2017. What's Next for Feminism? Presented by Patricia Karvelas. Radio program *The Drawing Room*, Produced by B. Wang. Broadcast 14 March 2017.

Rigby, Kate. 1998. Women and Nature Revisited: Ecofeminist Reconfigurations of an Old Association. *Arena Journal* 12: 143–169.

Rose, Deborah Bird. 1995. Women and Land Claims. In *Land, Rights, Laws: Issues of Native Title*, 1–8. Canberra: Native Title Research Unit, Australian Institute of Aboriginal and Torres Strait Islander Studies.

Ruether, Rosemary Radford. 1975. *New Woman, New Earth*. Minneapolis: Seabury Press.

Salleh, Ariel. 1984a. Contribution to the Critique of Political Epistemology. *Thesis Eleven* 8: 23–43.

———. 1984b. Deeper Than Deep Ecology: The Eco-Feminist Connection. *Environmental Ethics* 6 (Winter): 339–345.

———. 2014. Foreword. In *Ecofeminism*, ed. M. Mies and V. Shiva. London and New York: Zed Books.

Spongberg, Mary. 1993. If She's So Great, How Come So Many Pigs Dig Her? Germaine Greer and the Malestream Press. *Women's History Review* 2 (3): 407–419.

Taylor, Anthea, Maryanne Dever, and Lisa Adkins. 2016. Greer Now: Editorial. *Australian Feminist Studies* 31 (87): 1–6.

Warren, Karen J. 1987. Feminism and Ecology: Making Connections. *Environmental Ethics* 9 (1): 3–20.

———. 1994. Introduction. In *Ecological Feminism*, ed. K.J. Warren. London and New York: Routledge.

Lara Stevens holds a PhD in Theatre and Performance Studies from the University of Melbourne. She is the author of *Anti-War Theatre After Brecht: Dialectical Aesthetics in the Twenty-First Century* (Palgrave Macmillan 2016), and editor and translator of *Politics, Ethics and Performance: Hélène Cixous and the Théâtre du Soleil* (Re.press 2016), essays by French feminist philosopher Hélène Cixous. In 2014 she was the Hugh Williamson Postdoctoral Fellow in the Australian Centre at the University of Melbourne.

8

Climate Guardian Angels: Feminist Ecology and the Activist Tradition

Denise Varney

Traces of feminist environmental activism date back to Romanticism but are diverse, intermittent, and less well known than other streams of feminist thought. Mary Shelley's novel *The Last Man,* published in 1826, in which the author imagines a catastrophic future blighted by deadly diseases, earthquakes, storms, and floods, all attributable to human neglect, is an early literary intervention into environmental concerns about the future. In the twentieth century, Rachel Carson's *Silent Spring* (1962), a book said to have 'substantially altered the course of history', exposed the damaging effects of pesticides, especially DDT, on the biosphere (Griswold 2012). In the 1970s, feminists brought the peace movement together with protection of the environment at two well-known sites: Greenham Common Royal Air Force base in Berkshire, United Kingdom, where women held an anti-nuclear missile activist camp, and at Pine Gap near Alice Springs, Australia, where the Women's Peace Camp of 1983 targeted a US-Australian military installation. Françoise d'Eaubonne coined the

D. Varney (✉)
The University of Melbourne, Melbourne, VIC, Australia

L. Stevens et al. (eds.), *Feminist Ecologies*,
https://doi.org/10.1007/978-3-319-64385-4_8

term ecofeminism in 1974 to describe the feminist ecological project as having two aims: the abolition of patriarchy and the establishment of a relationship with the environment (see Rigby 2001: 27). Australian feminist philosophers Freya Mathews, Val Plumwood, and Ariel Salleh advanced critiques of the unacceptable anthropocentrism and androcentrism of capitalist patriarchal systems in the West. Here traditional feminist exposure of the domination of Women in patriarchal societies extends to the domination of Nature with both having their roots in entrenched cultural practices underpinned by philosophical and theological traditions.

This chapter updates the scholarship on feminist environmentalist activism with a study of the Australian-based Climate Guardians, whose performance-based protests take the form of silent public vigils for a safe climate. Whereas the all-female groups Guerrilla Girls, Pussy Riot, and Femen, for example, are the subjects of studies in activism against the patriarchal basis of art institutions, religion, and neoliberalism (Rosenberg 2016; Diamond et al. 2017), ecofeminist performance remains a minority form of both feminist and environmental activism. The Climate Guardians' stated aim is to place the environment centre stage in the interests of highlighting 'the vital role of guardianship of precious natural resources, both human and non-human, in addressing the global threat from climate change'.[1] The group's activism points to a reinvigorated environmental protest movement that, while attracting mainstream media attention, allows for a type of political 'infecting' that slips around authoritarian obstacles to conventional protest.

Their most compelling strategy of slippage is that this multi-age, volunteer ensemble of women and some men, numbering at times over 100, perform actions dressed as angels. They include cultural feminists, peace and harmony advocates, Greens activists, artists, academic feminists, historians, scientists, psychologists, and teachers in an assemblage of mixed ideologies, occupations, and backgrounds. They intercede in contested environmental spaces and prominent scenes of international climate diplomacy, gathering in mainly outdoor, site-specific spaces to enact techniques of durational passive resistance to climate change inaction. The aim is to create the kind of public spectacle that shames corporations and lawmakers for their ongoing failure to address pending ecological crisis, and incite public criticism. At their blockade of the G20 2014

Leaders' Summit in Brisbane, for example, they declared: 'our Prime Minister turned his back on the future by declaring his loyalty to coal' (ClimActs 2014). Performances might typically end with the arrest of individual angels, captured on film and then circulated as a further call to arms on social media. The Climate Guardians' appropriation of the angel figure and their live presence means they stand out in the modern cityscape and do much to reverse the diminution of women and nature in the public sphere while highlighting human dependence on 'biospheric processes' (Plumwood 2003: 21).

To analyze the Climate Guardians' actions, I turn to Plumwood and Salleh to discuss the radical feminist underpinnings of the Climate Guardians, and the question of woman, nature, and environment, to explore both the symmetry and differences between the feminist and environmentalist movements. The history of the angel as a religious figure raises further questions about harnessing spirituality to address environmental issues endemic to modern life, and for guidance on this matter I turn to Rigby's essay 'Women and Nature Revisited' (1998). From Mathews I consider the ecofeminist elements of the Climate Guardians' performances focusing on their appearances at the United Nations' conference on climate change in Paris in November 2015. I conclude by noting how the Climate Guardians offer a powerful mode of environmental activism based in techniques of silence and presence that express solidarity with human and non-human nature.

Ecology Environment Nature

Feminists have pointed to the oppressions inherent in binary value systems that divide the human from the non-human, humanity from nature, culture from nature, and man from woman, where the non-human, nature, and woman are tied to each other as secondary terms. As Rigby points out, Simone de Beauvoir argued that 'the association of woman and nature was a key element in patriarchal ideology that served to legitimate women's exclusion from the public sphere and their confinement to the home' (2001: 28). If as feminists argue Woman is a construct, Nature, as a proper noun, is similarly a construct with a history and a relationship to cultural tradition and knowledge formations.[2] Ariel Salleh's early work drew on the writings of Mary Daly and Adrienne Rich to critique 'the embeddedness of western sex stereotypes in the nature-culture dichotomy' in which Woman is an 'alleged natural being' inferior to and exploited by capitalist life processes (Salleh 1981: 8). In her theoretically masterful book *Ecofeminism as Politics: Nature, Marx and the Postmodern* published in 1997, Salleh established a mode of transdisciplinary scholarship that sought to destabilize the Western construct of humanity and nature as separate spheres. In her critique of Marxism and Frankfurt School social theory, she wrote, 'Nature remains the passive, unspoken substrate of theoretical subsumption, an exercise legitimated by the hardheaded patriarchal dichotomies of fact and value, nature and culture' (4). Here nature provided the resources that fueled the Industrial Revolution, but its materiality or agency was not to be recognized until much later. Yet second-wave feminism was at best ambivalent about nature and primarily concerned with women's rights. In her influential book, *Feminism and the Mastery of Nature*, first published in 1993, Val Plumwood identified this Marxist and second-wave feminist blind spot when she wrote:

> A feminist account of the domination of nature presents an essential but difficult further frontier for feminist theory, all the more testing and controversial because the problematic of nature has been so closely interwoven with that of gender. (2003: 1)

The interweaving of woman and nature as jointly subjugated under capitalism, socialism, and patriarchy found itself caught up in a negative bind in some early accounts of ecofeminism. The following critical proposition is one such example:

> Ecofeminism has developed, and continues to focus on developing, a body of complex theory in its attempts to explain and act upon the interconnected subjugations of women, other humans, and nonhuman nature. (Mack Canty 2004: 175)

Plumwood's more enabling solution was to offer an affirmative formulation that opened up a space for ecofeminism within the project of liberation. Here she posits that gender, race, class, and nature are 'the four tectonic plates of liberation theory', which if they acted in unison could 'shake the foundations of our conceptual structures' (2003: 1). For Plumwood, nature is 'a political category' marked by contested values and practices and includes the rights of nature, which was a groundbreaking assertion in the 1980s. On this view, ecofeminism is the movement in which feminists bring their friend nature to the political table.

Since Plumwood's death in 2008, the frameworks for thinking about climate change and its impact on human and non-human life have had to expand to respond to the increasing urgency of critical levels of global warming, rising sea levels, and the increasing frequency and intensity of fatal bush and forest fires, and floods and storms. Science confirms that global warming is the result of anthropogenic greenhouse gas or carbon emissions, the overuse of fossil fuels, and deforestation. But at the political level, inaction on climate change in Australia, as Robyn Eckersley points out, is at a 'political impasse arising from deep political polarization', which puts 'a bipartisan response to climate change out of the political reach of the legislature' (2015: 140). Hence the conditions are in place for extra-parliamentary ecofeminist activism.

Climate Guardians

Tasmanian environmentalist Allana Beltran initiated the first 'documented' intervention in Australian politics by an 'angel' in a performance installation known as the Weld Angel. Protesting the logging of old growth forest, she appeared in the Weld Valley, Tasmania, in 2007 harnessed to vertical and crossed saplings in a durational performance utilizing presence and silence that lasted 10 hours. The photographs show a woman in white face with red lipstick dressed in a long white dress with large handmade feathered wings against a background of forest and sky

(Beltran 2017). The performance became widely known when Beltran was sued by Forestry Tasmania and the Tasmanian Police for police time and wages costs in a case that ran for six months before being dismissed. Consulting closely with Beltran, Liz Conor, and Deborah Hart in Melbourne adopted the angel iconography in 2013 for climate change action, extending Beltran's concept of solo action to call for a multitude of angels, who would gather at predetermined sites and become known as the Climate Guardians.

Conor is an academic, writer, and activist whose previous campaigns included The Mothers of Intervention, who campaigned for maternity leave, and the John Howard Ladies Auxiliary Fanclub, which consisted of four 1950s housewives who popped up at public events to cheer then Australian Prime Minister, John Howard, in order to protest his outdated social and moral values. Hart, also a writer and activist with a background in arts management and fundraising, co-founded CLIMARTE: Arts for a Safe Climate in 2010 and remains on its board. She and Conor founded the independent network ClimActs: Acting for Climate Justice, of which the Climate Guardians was the founding act. ClimActs' website hosts a living archive of actions from 2013 to the present (ClimActs). Activists who join Climate Guardian actions might also join other troupes including the Coal Diggers, a billionaire coal miner's fan club, the medievalists of the Flat Earth Institute, the Frackers' Guild, or the recently formed Hackers Guild that specializes 'in mining and psychoanalyzing unsuspecting peoples' personal data and social media profiles in order to manipulate their minds and votes!' (ClimActs 2017a).

Conor and Hart as experienced, well-networked, and inventive activists developed smart ecopolitics for the contemporary era, playing humorously with the signifiers of historic oppression and subjugation of women and the environment without succumbing to an enervating paralysis or hard-edged didacticism. Each performance is documented and underpinned by extensive writings, which are archived on the ClimActs' website, that discuss the climate politics at stake in each chosen location.

The first Climate Guardian actions in early 2013 involved what they referred to as visitations wherein they visited the headquarters of BHP Billiton and the big four Australian Banks—CommBank, ANZ, NAB, and Westpac—during peak hour lunchtimes to stencil the Climate Guardian logo onto the buildings' glass doors and marble walls. They communicated through the action of being present in the space, acting in unison, and through the words and image of the logo, while remaining silent. The logo incorporated the words Climate Justice encircled by baroque-style angel wings, which remained, until workmen were sent to erase it, as a stigmata on the corporations' public interface drawing attention to its body politics. In the time-honored tradition of activist and avant-garde movements, the logo was an important materialization of the Climate Guardians' manifesto, entitled Our Safe Climate Demands. It states:

1. The paramount duty of government is to protect its citizens from grave threats;
2. Rapidly accelerating anthropogenic climate change places the very future of human civilization and the ecosystems upon which it depends at dire risk;
3. Urgent action is required to avoid further damage and to restore a safe climate;
4. The necessary action will require society-wide mobilization of resources at a scale and speed never before seen in peacetime, failure is not an option (ClimActs 2017b: 'Safe Climate Demands').

Since then, the ensemble's mode of performance has drawn on angel iconography, bodies in space, and textual signage. Actions include visitations, which consist of hosting, gathering, and manifesting en masse, or in small assemblies, in public spaces, and scenes of climate politics. A typical Climate Guardian performance consists of arriving at a location, gathering, harnessing the wings to the body, and then silently standing or walking, sometimes carrying placards. A key tactic is for the ensemble to assemble for a publicity shot in proximity to an iconic symbol of the capitalist-industrial-carbon producing order such as banks, mining companies, and politicians' offices. In 2014, seven Climate Guardian Angels performed a spectacular living sculpture, entitled 'Coal Requiem', at the Lorne Sculpture Biennale on the Great Ocean Road in Victoria. Commissioned for the opening of the three-week festival, the work was performed on the foreshore with angels elevated against the evening sky, accompanied by a female singer backed by two young brothers playing cello and violin. The words 'Coal Requiem' were written on the sand with burnt wood (resembling coal) and seaweed. In addition to the haunting songs, the classical musicians performed a nature-honoring piece they had specially composed for the occasion. The performance concluded with the angels igniting rescue flares to draw attention to the effects of carbon emissions on rising sea levels, especially in the Pacific region (Hart 2015: 68). Among their many appearances since then, the Climate Guardians have blockaded the entrances of corporate mining headquarters, Parliament House of Australia, political party buildings, and politicians' offices.

Regional visitations have included the dusty road into mining company Santos' Leewood's wastewater treatment facility in the Pilliga Forest. There the arrest of a Climate Guardian created a stunning ironic media image, which is archived on the ClimActs' website, that depicts a policewoman arresting an angel. The intersection of the sitting female angel figure, a vision of white robes and ethereal presence, and a standing female police officer upholding the laws of the state makes a powerful Brechtian gestus—a theatrical term for physical action, mode of speaking or facial expression that reveals a 'highly complicated and contradictory' relationship between characters in a play (Brecht 1984: 198). Here it exposes the relationship between the mining, fracking, carbon-producing economy, and the government that serves and protects it. The image of the Climate Guardian pleading with the policewoman for the right to make a peaceful protest amply displays the limits of the democratic state. The placating hand of the well-drilled female policewoman, who (literally) stands for the collusion of mining and the conservative government of New South Wales, is the gestic signifier of its entwined interests. Here the Guardians modernize feminist traditions of peaceful environmental activism honed at Pine Gap Women's Peace Camp and Greenham Common for the era of global media (see Chap. 8).

The Guardians are also aware of the combined effects of performativity and theatricality. Gathering in New South Wales in May 2016, during a mass civilian blockade against the export of coal to China and India, the Climate Guardians joined more than 2000 protestors to stage an Australian action in collaboration with the global #Breakfree from fossil fuel movement. Dressed in their customary white robes and angel wings, several Climate Guardians lay on the tracks of a railway bridge where they were arrested with 66 others for effectively disabling the coal trains passage to Newcastle Harbor, Australia, which is considered one of Australia's largest coal export ports (Hart 2017; *Newcastle Herald* 2016).

While Climate Guardian performances are visually striking interventions in the politics of climate change, their quiet evocation of the angel figure offers an intriguing constellation of cultural and religious iconography. Deploying elements of mimesis and parody, they offer complex interpretive conundrums. They appear pious and compliant at the same time as they are warriors for climate change, claiming agency as activists and extending this agency to environmental sites such as oceans, forests, and city buildings. Each of these physical actions symbolically speaks back to the power of corporate interests. And while the spectacle of women representing angels in long white dresses harnessed to massive organza wings appears to reinstate an idealized female figuration, the appropriation of the traditionally masculine identity of the angel, as warrior and harbinger, is self-reflexively performative. Designed for optimum media exposure, the actions of the passive-resistant angels manipulate the multiple signifiers they evoke.

COP 21 Paris

The Paris Climate Change Conference in December 2015, known as COP 21, was an important international platform for the Climate Guardians and involved, of necessity, navigating the links between hard politics and politically motivated art. The Paris conference was the twenty-first meeting of the nation-based Parties to the Kyoto Protocol, an internationally binding agreement reached in Japan in 1997 to set carbon emission reduction targets, although a number of governments had not signed it. Australia under the conservative Government of John Howard refused to be a party to the agreement, and it was not until the Labor Government under Prime Minister Kevin Rudd came to power in 2007 that it signed onto the agreement in line with the new Prime Minister's earlier declaration that climate change was 'the great moral challenge of our generation' (Rudd 2007). The politics of climate change came to a head in 2012 with a major electoral loss for the Labor Government and the return of a far right-wing coalition under Prime Minister Tony Abbott, which quickly repealed laws designed to reduce carbon emissions. The Climate Guardians flocked to Canberra to hold a press conference at Parliament House to mark the occasion and to remind politicians—as their banner stated—that it was 'Five Minutes to Midnight' on the Doomsday Clock. Since then Australia has arrived, as Eckersley puts it, at the 'political impasse' that the Climate Guardians would seek to remedy in Paris (2015: 140). Attended by 150 Heads of State and over 25,000 accredited delegates, the United Nations conference also held A Global Festival of Cultural Activity on Climate Change, an extensive multi-arts event that ran from September to December 2015 involving artists from all around the world. This structure enabled seven Australian Climate Guardians to attend the conference as artists engaged in the cultural festival.

COP 21 would take place in the wake of the Paris terror attacks of 13 November, in which 130 people were massacred and hundreds wounded. As a consequence of raised security, a State of Emergency was in force in the city banning public gatherings and street protests for the duration of the conference and after. This ban was potentially disastrous for the Climate Guardians' outdoor site-specific actions. However, by identifying as artists engaged in cultural activity, they were permitted to perform visitations at the Eiffel Tower and the Louvre (Fig. 8.1).

The Guardians' actions were political in the context of the Australian Government's inaction on global warming. They mounted a blockade outside the offices of the French multinational energy company Engie, owner of the high-polluting, brown coal-fueled Hazelwood Power Station

Fig. 8.1 Climate Guardians at the Eiffel Tower, Paris, 2015 (Courtesy of the photographer: Maggie Miles)

in the Latrobe Valley in Victoria and its adjoining opencut mine. A Climate Guardian Press Release, intended for its Australian followers but available to a global audience, announced that 'A Flock of Climate

Guardian Angels from Australia descended on Engie's global head office in Paris to decry the corporation's appalling treatment of Victoria's Latrobe Valley community and their environment' (ClimActs 2015). (Since decommissioned, the power station closed in March 2017, but site remediation will take many years at costs of nearly three quarters of a billion dollars.) The Climate Guardians' Paris schedule included the delivery of primary school children's letters and art works to Laurence Tubiana, French Ambassador for COP 21. Unable to make contact with Tubiana, the letters were delivered instead to Australian Greens' Senator Larissa Waters, posted online, and read out to delegates as they arrived at the conference centre at Le Bourget on the opening day of the UN COP 21 climate talks (ClimActs 2015).

Liz Conor, one of the two conveners of the Climate Guardians, told the author that the performances in Paris, so soon after the terror attacks in the city, took on a wider significance than they had anticipated and for which they were unprepared (Conor 2016). Their first outing was to Place de la République on 29 November to visit the Marching Shoes Installation, organized by Avaaz (a global activist network). The installation consisted of thousands of pairs of shoes that had been placed in the square by Parisians and visitors in lieu of the Climate Change March that was canceled due to the State of Emergency (Hart 2017). An image of the Climate Guardians walking among the shoes appeared on the front page of *Le Monde* online, after which they were widely recognized as they passed through the streets. Hart also recalls that 'the response was more powerful than we had envisaged' (2017). Here religion, climate politics, terrorism, and grief came together in ways that showed how interconnected issues can be as they flow around pivotal points such as the appearance of the Climate Guardians and their silent empathetic presence in the city. The Climate Guardians' final appearance at COP 21 was to hold the Red Line, which involved angels walking in a horizontal line holding a length of red cloth to signify 'the boundary, the limits beyond which both the climate and the biosphere as well as our social systems collapse' (Hart 2017). On this twelfth and final day of the conference, they participated in a march, despite the restrictions, in a more militant mode of activism and were joined by other groups.

Angels in Religion and the Arts

The response of Parisians reveals the final conundrum of the relationship between the Climate Guardians' contemporary embodiment of the angel figure and the angel whose origins traverse the three great monotheistic religions of Judaism, Christianity, and Islam. In these patriarchal traditions, in which the angel is an emissary of God, a warrior, and spiritual being, he is unquestionably a masculine figure, whose capacity for transcendence is central to the role of mediator between the exalted deity and the human world. However, when Biblical and New Testament versions of the angel were transformed into visual imagery by Christian artists in Rome around the fourth century, artisans drew on Hellenistic representations of the Winged Goddess of Samothrace (Nike), which takes a feminine form (Jones 2011: 16). (The Winged Goddess is on display in the Louvre outside which the Climate Guardians assembled at one of their daily appearances.) Thereafter, the Christian angel had wings and a certain gender ambiguity that remains today. Central to the Christian tradition is the story of the Annunciation in which God sends the Angel Gabriel to persuade the Virgin Mary to submit to the Holy Ghost, who would 'cover' her womb after which she would give birth to Jesus, the Son of God (Luke 1: 26–38). The angel facilitated the central miracle of the virgin birth, in which the Christian God would assert moral superiority over the pagan Zeus, for example, who took on the form of a swan with beating wings to rape the maiden Leda, as in the myth of Leda and the Swan. Leonardo da Vinci's 'The Annunciation', painted 1472–1475, represents the angel as a young Renaissance man, civilized and brilliantly robed, kneeling in an attitude of respect before the Virgin Mary, who is seated in blue robes. His feathered wings rise from his shoulder blades like two Nike arrows. This painting gives the angel a humanist form that is aristocratically beautiful, and persuasive, as befits his role as God's messenger, while Mary remains passive in the form of the vessel she will embody.

Angels were feminized in Western European visual arts in the nineteenth century when they came to be associated in the late Romantic imagination with the non-rational. As Michelle Le Doeuff notes drolly, once a category is feminized, then it is devalued, and so by the time angels appear as women, they have lost most of their theological, philosophical, and intellectual associations: 'at best the relationship to knowledge proposed for women is precisely the one some men no longer want' (2003: 7). Victorian England gave rise to Coventry Patmore's poem 'The Angel in the House' (1854–1863), which represented an idealized femininity associated with servile roles as carers, nurturers, and attendants. As Jeannette King writes, the angel in the house was a 'sexless angel' who crossed into 'domestic ideology, embodying all the Christian virtues of love, purity and self-sacrifice so as to act as moral centre of the family' (2005: 11). At the same time, a more assertive representation of the female angel emerged in female aerialists in theatre and in the circus, who, as Peta Tait explains, commonly used angel motifs in their acts (2005: 19).

Angels have returned as potent symbols in twentieth-century literature, theatre, and film. Benjamin's much quoted reading of Paul Klee's 'Angelus Novus', painted in 1920, is of a male Angel of History caught in the storm of modernity's destructive force. His 'eyes are staring, [his] mouth is open, his wings are spread' in a gesture that looks back at catastrophe in a way that we might interpret today as impending ecological disaster (1999: 249). Wim Wenders brought melancholic angels to Berlin to oversee the dying days of socialism in the 1980s' film *Wings of Desire*, where they are, according to tradition, gendered male with orderly white wings protruding from dark overcoats. Here they watch over epochal political change aligning the angel entity with ruptures in modern history, as witnesses, notetakers, and as a counterforce of calm, meditation, and silence. Tait reminds us too that Bruno Ganz's angel 'loves a female aerialist who wears white (chicken) wings' (146). Drawing on the nineteenth-century female aerialists' uses of the angel motif, Angela Carter's *Nights at the Circus* (1984), a novel set in a traveling circus, gave angel wings to the athletic and daring Sophie Fevvers, who used them to forge a career as a famous aerialist. Her winged victory over other aerialists elevated her to celebrity status. Unlike Wenders' invisible to the human eye melancholic angels, flamboyant wings are also crucial to the female Climate Guardian Angels because they help to constitute performative figurations of a human, non-human, animal hybrid, a spectacular materialization of climate change critique, and a compelling manifestation of the judgments of history. Tony Kushner's spectacular theatrical work, *Angels in America: A Gay Fantasia on National Themes* (2017), was first performed in 1991. Lara Stevens draws attention to the role of the 'glittering angels' both male and female who crashed 'through New York apartment and hospital ceilings to deliver messages to sick mortals' on political and philosophical matters to do with the American legal and political system (2016: 51).

In modern terms angels can be said to occupy a theological, extra-judicial, extra-political, and more-than-human world that accords them the possibility of a degree of impunity from human juridical and political interventions. Their more-than-human form, their capacity to appear, to be selectively visible to both victims and the blessed, and to fly and swoop places them symbolically within the heavens, and within the natural world of landscapes, waterways, and clouds, and in ancient and modern cities. Like metaphysical and mythical beings, and figures from folktale, in the visual arts they are also depicted in interstitial spaces such as stairways, patios, entrances, and dungeons signifying the mediating role they play as messengers rather than protagonists.

Ecofeminist Performance Activism

Embodying the simultaneous domestication of the angel as a passive female and its expression of female autonomy, the Climate Guardians appear graceful, even passive, but their techniques are designed to wield maximum power across contemporary live and digital platforms. Beltran's powerful and overtly feminine Weld Angel defied the angel figure's masculinist and religious past reassigning it as an ecofeminist warrior, a messenger, and a more-than-human figure with feathered wings and wooden poles attached to a body. The Climate Guardians, like Beltran, draw on feminine traditions not to reactivate them in a conservative way but to dissolve the divisions between the human and non-human worlds, to reject dualisms even as they evoke them. They counter that which Sherilyn MacGregor refers to critically as the 'masculinization of environmentalism' (MacGregor 2010: 230) and use strategically designed iconography to make powerful points about the global fossil fuel economy.

The modes of performance favored by the Climate Guardians are to perform in unison, adopt angel-like behavior such as mediation, and perform a non-individuated embodied self. The host of angels offers an alternative to the dangerous individuation of the post-Enlightenment world. By performatively rejecting individuation, their collective action discards both leftist identity politics and neoliberal individualism. Their humans-with-wings appearance unsettles the dualism of the human and non-human, aligning with Plumwood and others who called for actions that promote the coexistence of the human, non-human, and object or material world. Their appropriation of the angel figure and their live presence means they stand out in the modern cityscape and do much to reverse what Plumwood refers to as 'backgrounding', a process that denies presence to the actions of women and nature in the public sphere (21).

They also activate the hallmarks of ecofeminism that Freya Mathews associates with the portrayal of humans in the natural world as a community of beings, related in the manner of a family, but nevertheless distinct. She goes on to argue that ecofeminist encounters between beings are 'open-minded and attentive', characterized by an attitude of care or compassion which can provide the grounds for an ecological ethic in an open field of enquiry (1994: 162). Mathews continues that 'concern for Nature is the product of a re-awakening to our kinship with our individual non-human relatives; … and it springs not from a "cosmic identification" with Nature but a sense of solidarity with our fellow beings' (162). This aspect of ecofeminism is evident in how, for example, rather than acting solely on their own behalf in Paris, the Climate Guardians delivered children's letters, having first engaged with teachers and schools in Australia. And in a further instance of solidarity, and an apparent departure from environmental interests, Parisian citizens recovering from recent terrorist attacks in Paris responded emotionally and politically to the seven Climate Guardians standing silently in public spaces at the Louvre and at the Eiffel Tower. The contemporary version of ecofeminism, as exemplified in this example, points to a convergence of feminism and ecology with new threats to do with terrorism and, by extension, advanced forms of chemical and fossil fuel biowarfare doing harm to humans, animals, lands, and the atmosphere.

There is a compelling logic to the Climate Guardian Angels' activism. Their demeanor is dignified, restrained, heavily codified, and disciplined. It is mostly not engaged in overt dissent or conflict. They assemble, stand, and walk mostly in silence. Their impact lies in their unexpected appearance in the modern secular world and the way they turn traditional gender associations of woman and nature into ecological activism and radical guardianship. Their effectiveness, if we accept the thesis of the post-political present, is to be ineffective politically. They trouble politicians and corporations by highlighting the contradictions of political inaction over climate change without resorting to a 'message based' didactic ecological art. Here the idea of guardianship of land and resources is placed in contradistinction to modern democratic systems that fail to find the means to deal with climate change. I hesitate to say this but their Gandhian style of peaceful protest carries the suggestion of the transcendent—given that their power also works through proximity, presence, and silence.

Notes

1. All references to the Climate Guardians' statements, press releases, and commentaries are from the ClimActs website: www.climacts.org.au. Discussion of the Climate Guardians' protests draws on photographic documentation of activist interventions that are also available at the same address. Where possible I have added titles to guide readers to the source. Elsewhere the site hosts an online archive where actions can be located by month and date.
2. Nature with a small 'n' refers to the 'the phenomena of the physical world collectively, including plants, animals, and the landscape, as opposed to humans or human creations' (OED 2017). Yet this innocuous definition reproduces uncritically the dualism of human and non-human beings and objects.

Bibliography

Beltran, Allana. Accessed 7 April 2017. http://www.allanabeltran.com

Benjamin, Walter. 1999. *Illuminations*. London: Pimlico.

Brecht, Bertolt. 1984. *Brecht on Theatre*. Trans. John Willett. London: Methuen.

Carson, Rachel. 1962. *Silent Spring*. Boston: Houghton Mifflin and Cambridge, MA: Riverside Press.

Carter, Angela. 1984. *Nights at the Circus*. London: Chatto and Windus.

ClimAct. 2014. G20 Climate Guardian Angels. 17 November. Accessed 19 April 2017. https://climacts.org.au

———. 2015. Updated: Climate Guardians Engie HQ Visitation. 10 December. Accessed 19 April 2017. https://climacts.org.au/2015/12/10/climate-guardians-engie-visitation/

———. 2017a. Hey Malcolm, Choose Our Hackers Guild! 4 April. Accessed 19 April 2017. https://climacts.org.au/climate-guardians/

Conor, Liz. 2016. Interview with author, 10 July.

Diamond, Elin, Denise Varney, and Candice Amich, eds. 2017. *Performance, Feminism and Affect in Neoliberal Times*. London: Palgrave Macmillan.

Eckersley, Robyn. 2015. Australian Democracy and Climate Politics for the Long Term. *Meanjin* 74 (3, Spring): 140–146.

Griswold, Eliza. 2012. How Silent Spring Ignited the Environmental Movement. *The New York Times Magazine*, 21 September. Accessed 19 April 2017. http://www.nytimes.com/2012/09/23/magazine/how-silent-spring-ignited-the-environmental-movement.html

Hart, Deb. 2015. *Guarding Eden: Champions of Climate Action*. Melbourne: Allen and Unwin.

———. 2017. Interview with author, 8 May.

Jones, David. 2011. *Angels: A Very Short Introduction*. Oxford: Oxford University Press.

King, Jeanette. 2005. *The Victorian Woman Question in Contemporary Feminist Fiction*. Basingstoke and New York: Palgrave Macmillan.

Kushner, Tony. 2017. *Angels in America: A Gay Fantasia on National Themes*. London: Nick Hern Books Ltd.

Le Doeuff, Michelle. 2003. *The Sex of Knowing*. Trans. Kathryn Hamer and Lorraine Code. New York: Routledge.

MacGregor, Sherilyn. 2010. 'Gender and Climate Change': From Impacts to Discourses. *Journal of the Indian Ocean Region* 6 (2): 223–238.

———. 2014. Only Resist: Feminist Ecological Citizenship and the Post-politics of Climate Change. *Hypatia* 29 (3): 617–633.

Mack Canty, Colleen. 2004. Third-Wave Feminism Reweaving the Nature/Culture Duality. *NWSA Journal* 16 (3): 154–179.
Mathews, Freya. 1994. Relating to Nature: Deep Ecology or Ecofeminism? *The Trumpeter* 11 (4): 159–178.
Newcastle Herald. 2016. Editorial: An Election Announcement in Canberra, a Climate Protest in Newcastle. 8 May. Accessed 21 May 2017. http://www.theherald.com.au/story/3894622/climate-guardians-take-hold-of-the-harbour/
Plumwood, Val. 2003. *Feminism and the Mastery of Nature*. London and New York: Routledge.
Rigby, Kate. 1998. Women and Nature Revisited: Ecofeminist Reconfigurations Of An Old Association. *Arena Journal* 12: 143–169.
———. 2001. The Goddess Returns: Ecofeminist Reconfigurations of Gender, Nature, and the Sacred. In *Feminist Poetics of the Sacred: Creative Suspicions*, ed. F. Devlin-Glass and L. McCredden, 23–54. Oxford: Oxford University Press.
Rosenberg, Tiina. 2016. *Don't Be Quiet, Start a Riot! Essays on Feminism and Performance*. Stockholm: Stockholm University Press.
Rudd, Kevin. 2007. Extract from Mr. Rudd's Introductory Address at the National Climate Change Summit. Saturday 31 March 2007. Posted on YouTube 6 August, 2007. Accessed 29 April 2017. http://trove.nla.gov.au/version/50805471
Salleh, Ariel. 1981. Of Portnoy's Complaint and Feminist Problematics: A Reconciliation with Critical Theory. *A.N.Z.J.S.* 17 (1): 4–13.
Shelley, Mary. 1985. *The Last Man*. London: Hogarth Press.
Stevens, Lara. 2016. *Anti-War Theatre After Brecht: Dialectical Aesthetics in the Twenty-First Century*. London: Palgrave Macmillan.
Tait, Peta. 2005. *Circus Bodies: Identity in Aerial Performance*. London: Routledge.
Wings of Desire. 1987. Dir. Wim Wenders, Argos Films.

Denise Varney is Professor of Theatre Studies in the School of Culture and Communication at the University of Melbourne, and co-director of the Australian Centre. She was co-convenor of the Feminist Research Working Group (IFTR) from 2011 to 2015. She publishes on Australian Theatre, Brechtian and contemporary German theatre, and feminist criticism and performance. Her latest book is *Performance Feminism and Affect in Neoliberal Times* (Palgrave Macmillan 2017), co-edited with Elin Diamond and Candice Amich. She is co-author of *Theatre in the Asia Pacific: Regional Modernities in the Global Era* (Palgrave Macmillan 2013) and is working on an ARC project, 'Patrick White and Australian Theatrical Modernism'.

9

Thinking–Feminism–Place: Situating the 1980s Australian Women's Peace Camps

Alison Bartlett

This chapter connects thinking, feminism, and place, to suggest a link between ecology and epistemology, between the environment in which we live our lives and the production of knowledge. Living in Australia has a particular impact on thinking, I suggest, while also locating thinkers in particular ways to what has become known as the 'Global North'. In order to think about time, place, and thinking as an ecological environment, I focus this chapter on the 1980s as particularly generative in the field of feminist thinking and activism.

The reflexive concepts of situated knowledge and partial perspective are legacies of 1980s feminist thinking, and a direct response to Cold War politics and ensuing feminist peace activism. Connected to those ideas are the actions of the 1980s women's peace camps which were notable not only for their local actions as part of global concerns about nuclear proliferation and the potential devastation of nuclear war, but also for their critique of the connections between militarism, masculinity, and violence.

A. Bartlett (✉)
The University of Western Australia, Perth, WA, Australia

L. Stevens et al. (eds.), *Feminist Ecologies*,
https://doi.org/10.1007/978-3-319-64385-4_9

My research interest in the 1980s women's peace camps in Australia was prompted partially due to their absence in texts but has been driven by their continuing significance for thinking through the connections between feminism, ecology, epistemology, and the military industrial complex. At a time when both feminism and environmental studies were being recognized and formalized in universities as legitimated forms of knowledge, the 1980s can be considered foundational in laying down key concepts and texts for the generations of scholars since.

In an effort to highlight the place of thinking, the chapter is structured around particular sites of ecofeminist activism. The sites involve bodies of thought, as well as thinking bodies; they are singular occasions, longitudinal actions, and intellectual moments. The chapter therefore moves between places in the United States, Britain, Europe, and Australia to demonstrate a collective body of feminist thinking and their connections with each other and the times in which they are produced. There are two apparently contradictory forces happening in this proposal: that knowledge is generated in particular times and places through embodied experience and that ideas and analysis travel between bodies, times, and places. The result is, I suggest here, what I am calling an epistecology.

Thinking in Place

In arguing for the ways in which embodiment, ecologies, and epistemology are entangled, I firstly reflect on the conditions of producing my own research in Australia and then situate them in relation to the models of knowledge production inherited from elsewhere. This reflexive beginning sets the scene for tracking a movement of ideas while establishing their connections and commonalities. Situating the production of knowledge by specific bodies in particular places has been a fundamental tenet of feminist research methodology, challenging a priori assumptions that knowledge is produced by objective minds and is of universal application and importance. The assumptions implicit in the idea of objectivity were critiqued by feminists in the 1980s, with Donna Haraway summarizing the 'science question' thus:

> All Western cultural narratives about objectivity are allegories of the ideologies governing the relations of what we call mind and body, distance and responsibility. Feminist objectivity is about limited location and situated knowledge, not about transcendence and splitting of subject and object. (1988: 583)

Situated knowledge and partial perspective became key legacies of these debates that obligate feminists to position themselves in relation to their work, rather than assuming universal vision.

The Australian focus of this book on feminist ecologies prompts me to ask what difference it makes to be thinking in Australia, and how the conditions of knowledge production might differ from elsewhere. Since undertaking my doctoral research while living in the tropics of far north Queensland, I have been attendant to the relation between thinking rhythms and the weather. My research into continental feminist philosophies occasionally found me trying to literally inhabit such a world of thinking: I would seek Hélène Cixous by making my world into a faux French salon and emulate an imagined Luce Irigaray, eating croissants and chocolates, playing Grace Jones, reading high theory. There was a perverse pleasure in playing the Francophile in the heat and sweat of slippery bodies and lives lived in the tropics, which seemed contrarily un-European and not inductive to French feminist philosophical thought. The tropics, as Paul Sutter notes, is itself a European discourse that registers its unsuitability for those (white people) from the temperate (European) zones (2014: 178). However unsuitable, these conditions trained me for research, so now, when the mercury rises, I register in my body that it is my optimal time to think and write. Heat, for me, signals thinking time.

Turning attention from heat to light, the Australian landscape artist Barbara Bolt notes a similar shift in the meanings of light, which affect the conditions of her production of artwork in Kalgoorlie, Western Australia (2000). Since the Enlightenment, European light has been associated with knowledge, illumination, vision, and clarity, Bolt argues (2000: 204). In Australia however the sun's glare transforms light into something blinding, that can become fuzzy through heat haze, that makes us squint, and causes pterygiums and cataracts in our eyes that obscure our vision (Bolt 2000: 206). If we cannot use the light as a metaphor for knowledge, what happens when work is produced in such conditions? Is the work not enlightened, or does it exceed those legacies? These examples of heat and light suggest ways in which the material conditions of knowledge production might be made apparent, throwing into disarray those models we inherit from elsewhere. Through examples like these we can begin to interrogate the ways in which we occupy and think in localized spaces about how particular places and their effects on the body impact upon thought.

Another aspect that impacts the conditions of knowledge production is the lived politics of the moment. This summer while I write, I have been distracted from my usual research by involvement in a community movement protecting the destruction of 97 hectares of ecologically rare and endangered wetlands and ancient banksia woodlands in my suburb in outer Perth, Western Australia. It is currently being cleared for a highway for trucks known as Roe 8, a controversial extension of the Roe Highway that was drawn up 50 years ago when this area was not even a part of urban Perth, when bushland and wetlands were routinely cleared on the Perth coastal plain for urban development. It was before climate change was commonly discussed by everyday folk, and before we knew better. Living so close to the site, I am distracted by the dust being blown through my house by the bulldozing nearby, the physical dirtiness of my writing space no matter how much I clean, the dryness of dust in my mouth. I am distracted by this local protest movement, which reflects the legacy of what I am researching in another place and time and yet remains connected in striking ways.

These examples illustrate the collapsing together of epistemology and ecology into the same term, epistecology, which foregrounds the environment in which knowledge is produced. It might be reminiscent of Lorraine Code's suggestion that communities produce knowledge: that 'knowing subjects and their engagements in informal and formal practices of knowing are mutually constitutive' (2008: 188). Communities of reflexive feminist thinking reveal the possibilities of epistecology.

Baltimore, United States, 1984

Adrienne Rich was conscious of the place of writing when she was preparing for the conference 'Women, Feminist Identity and Society in the 1980s' held in Utrecht, Holland, in 1984. It was here that she presented her keynote address, 'Notes Toward a Politics of Location'. Troubled by the grand narratives of women's oppression, Rich wanted to discuss specificity. As a collective pronoun 'woman' can no longer seem to contain the differences between women and their dispersal in space, in conditions, histories, and politics. While writing her paper in the United States, she recalls a childhood game of addressing correspondence and thereby begins to situate herself:

> Adrienne Rich
> 14 Edgevale Road
> Baltimore, Maryland
> The United States of America
> The Continent of North America
> The Western Hemisphere
> The Earth
> The Solar System
> The Universe (Rich 1985: 8)

This form of address positions Rich within the universe, rendering her insignificant by virtue of its scale. It is both specific and universal, and suggestive of the micro- and macropolitics of location. Rich's paper discusses the conditions of American society in the early 1980s. She speaks of McCarthyism, the Iron Curtain, the threat of nuclear annihilation, militarism, and women's activism around the world and across the century. She also speaks of her domestic dwelling, the conditions of her writing, the bumblebee that gets caught inside the house, the honey jar on the bench, the books on her table. She talks of her white woman's body which locates her planes of privilege and oppression. This continually shifting scale connects the domestic and the worldly through politics and poetics, moving from: a body; a house; a conference; pharmaceutical corporations; poisoned rivers; atomic testing in deserts; urban hospitals closing; the first female astronaut; grape hyacinths wildly spreading in the garden; and cruise missiles being stockpiled in Greenham Common in Britain, Italy, Belgium, West Germany, and the Netherlands. This might seem like a random list, but it poetically demonstrates the connectedness of the intimate and global, the domestic and the political. Rich notes 'the valorization of manliness and masculinity. The armed forces as the extreme embodiment of the patriarchal family' (1985: 17), and '[t]he growing urgency that an anti-nuclear, anti-militarist movement must be a feminist movement, must be a socialist movement, must be an anti-racist, anti-imperialist movement' (17). It needs to be global.

The idea of epistecology used here might also remind us of Mieke Bal's travelling concepts: that 'while concepts are products of philosophy and tools of analysis, they are also embodiments of the cultural practices we seek to understand through them' (2002: 21). Concepts are anchored in the local at the level of embodiment even while referencing the global and travelling between other times and places.

Greenham Common, United Kingdom, 1981

By June 1984, as Rich reminds her audience, the US Government had already stationed nuclear missiles around Europe, including at Greenham Common in Berkshire, just 50 miles from London. The missiles are stationed at US Air Force bases which are part of the North Atlantic Treaty Organization (NATO) agreement after World War II in an effort to prevent any future nuclear world war. The immediate threat to the West was perceived to be Communist, specifically the Union of Soviet Socialist Republics (USSR). This legacy is apparent in the popularity of espionage film and fiction since that time (Shaw 2007). It was widely narrated in public discourse, as Shaw demonstrates in film and fiction, as an arms race with the West pitted against the USSR to develop more sophisticated nuclear technology. The development, testing, and then stationing of the newly tested Pershing and Cruise missiles in the 1980s by the United States marked a shift from preventative strategies to a more menacing potential for active nuclear strikes on the USSR. It also made

American missile locations around Europe potential targets for strikes from the USSR, which sparked growing public concern and activism from around 1980 (Roseneil 2000: 42).

In 1981 a group called Women for Life on Earth decided to walk the 110 miles from Cardiff in Wales to the US Air Force base at Greenham Common. Their walk aimed to generate media and public attention about the imminent installation of missiles there. They wanted a public debate about the decision which was made between Prime Minister Thatcher and American President Reagan. When they reached the base on 5 September 1981, some women chained themselves to the mesh fence. The base superintendent was informed of their concerns for the future of humanity, but he was unperturbed and told them to stay chained as long as they liked (Laware 2004: 20). So they stayed. They borrowed camping equipment and set up camps. They remained protesting until the missiles were delivered two years later, in 1983, and then until the missiles were removed in 1991 following the Intermediate-Range Nuclear Forces (INF) Treaty signed by American President Reagan and Russian President Gorbachev in 1987.

At the camp, numbers swelled at weekends, and when campaigns were organized to coincide with military activity (Laware 2004: 22). In December 1982, for example, 30,000 women converged to Embrace the Base, holding hands around the nine mile perimeter (Laware 2004: 21). In December 1983, after the first missiles were delivered, 50,000 women protested at the base. Hundreds camped at the various base camps around the perimeter (Roseneil 2000: 69). Tens of thousands of women visited the peace camp, which transformed lives and politics (Roseneil 2000: 69). In winter when the dogged English rain turned the common land to mud and then ice, they still stay. In 'the heart of the English countryside' the Greenham women contest the use of the Common as a military complex and transform it into a site of social protest (Cresswell 1996: 131). Greenham Common becomes forever associated with the women's peace camp. It becomes what Alain Badiou calls an 'evental site' (2005: 179), a site forever associated with a singular event that upsets the established order (Bartlett 2016: 7). It is iconized in exhibitions, books, memoirs, memorials and online exhibitions, local history walks, and national history. This is the anti-nuclear, anti-military feminist movement that Rich speaks about in 1984.

Situating Knowledge

In 1986, philosopher Elizabeth Grosz is thinking through a similar quandary to Rich. She wants to 'explain both the commonness women share cross-culturally, and their cultural and individual specificities' (1987: 2). The similarity is mirrored in the title of her article, 'Notes Towards a Corporeal Feminism', published in the new flagship journal, *Australian Feminist Studies* in 1987. The tentative title 'Notes Towards' belies the shift in thinking but suggests a fledgling disciplinary body of knowledge. For Grosz, the solution to situating knowledge is located in the body:

> Women's carnal existence, their corporeal commonness, may provide a universal 'raw material', which is nevertheless pliable enough to account for cultural, historical, class and racial specificities distinguishing concrete women from each other. (1987: 2)

The body is able to contain contradictions and difference but also account for privilege and oppression, and Grosz folds it into a psychoanalytic model of subjectivity. She later argues that '[t]he body functions as the repressed or disavowed condition of all knowledges' (1994: 20). The body becomes an epistemological site as well as the condition of the production of knowledge.

Grosz is thinking about how to uncouple the body from its Cartesian legacy, which neglects the body in favour of the mind in a series of binary oppositions. Women and men, nature and culture, private and public, emotion and rationality, passive and active, carnal and intellectual: all of these oppositions are hierarchized to privilege and align men with culture, the public sphere, rationality, activity, intellectual thought. These Enlightenment values are embedded in the beginnings of democracy, citizenship, civic identity, and scientific knowledge production. Rethinking the body, she argues, means rethinking subjectivity, and everything that rests on it, thus transforming those oppressive hierarchies. And so she goes on to outline possible directions for reimagining space, time, power, representational systems, and sexuality.

Kakadu National Park, Australia, February 1985

In late February 1985, at the start of the monsoon season, a white woman is kayaking alone in Kakadu National Park (Plumwood 2012: 10). The land of the Bininj/Mungguy people, this is a monumental landscape of sheer chasms and wide rivers, estuarine tributaries, and ancient vegetation in what is known as the stone country of Arnhem Land in the tropics of northern Australia. This woman often kayaks alone and is highly experienced, travelling to adventurous and thrilling places as she has this time to the remote National Park, where the film *Crocodile Dundee* is due to be shot a few months later. Out of the brackish river on the first day of the monsoon season, a crocodile's eyes emerge to gaze at the kayaker, then the animal leaps up to grab her by the legs and drag her under water in a death roll, and then another, and then another. Miraculously, she is released and drags herself up the banks to crawl in search of help. Her journey takes hours, slowed as she is by the sustained injuries. Not many people escape a saltwater crocodile's death roll, let alone three, but this experience of environmental philosopher Val Plumwood proved to be defining in her thinking. While Plumwood's articulation of the human/nature divide as a form of human chauvinism links those Cartesian dualities to systemic male-centred systems of thought, encountering the crocodile as predator redirected her to thinking about human beings as prey, as food for others. 'We are food', she writes, 'juicy, nourishing bodies. Yet, as I looked into the eye of the crocodile, I realized that my planning for this journey upriver had given insufficient attention to this important aspect of human life, to my own vulnerability as an edible, animal being' (2012: 10). Not only food, but meat: 'that my body, like theirs, was made of meat' (10). This fundamental reduction of embodiment as prey to others seeking food—'we are the feast' (15)—inverted Plumwood's understanding of human relations to the world and informed her thinking about ecological animism in which humans are just one part of the complex ecology of an agentic world.

Plumwood's encounter with a crocodile was transformational in her life and thinking, and symptomatic of the intimacy between the corporeal and the conceptual in feminist thinking. It can be understood in terms of Grosz's contemporaneous project: that the body is the 'disavowed condition of all knowledges' (1994: 20), except that for Plumwood it is not disavowed but in full sight and acknowledged.

Pine Gap, Australia, November 1983

On 15 November 1983 when the first Pershing and Cruise missiles were being delivered for storage at Greenham air base, one of the many supportive actions around the world was taking place in the middle of Australia at the US military base at Pine Gap, 20 kilometres from Alice Springs in central Australia, where it remains today. Women For Survival—an umbrella organization that aimed to bring together women's, peace, and anti-nuclear groups from around the country—organized a peace camp in the middle of the desert. Camping on the side of the bitumen road leading to the Base, around 800 women spent over two weeks there enduring blistering temperatures of 40 degrees on several days. At night the temperatures plummeted. They organized into 'affinity' groups, ran non-violent direct action training, shared skills for media and police liaison, and decorated the fence with banners made by those who could not be there. They mounted daily events for the media in order to bring attention to this little-known piece of Cold War architecture in the heart of Australia (Bartlett 2013: 917).

For some of the women from the city, it is their first contact with Aboriginal women from traditional homelands (Somerville 1999: 24). For some Aboriginal women it is their first contact with lesbians (Kelham 2013: 87). The Aboriginal women do not stay long, keen to avoid police provocation (Kelham 2013: 84). But one night there is traditional dancing by the Arrernte and Pitjantjatjara women with the white women joining in (Somerville 1999: 23). Telegrams of support arrive from Greenham Common, from unions, and from politicians. The camp makes the front page of *The Sydney Morning Herald* newspaper.

The Australian desert has an implicit appeal to the nation's metaphorical heartland, as its centre, the heart of Australia (Bartlett 2013: 918). Camping in this terrain is a measure of dedication for the largely urban protestors, but it is also a convergence of multiple meanings of place. Reflecting on her experience at the camp, Margaret Somerville articulates three layers of meanings:

> It is clear now that Aboriginal women did, in fact, present the first tissue paper layer of a different level of mapping the land with songs and songlines; that the disruption of straight lines and roads by circles of [women protestors'] hats and parasols opened up possibilities for new meanings to emerge and that those possibilities are still open. And my embodied three-dimensional mapping of hills and rivers, valleys and caves, is a sand map written on the body. (1999: 43)

Aboriginal song mapping, the collective action of protestors, and the embodied individual thus transform the meanings of place. Grosz considers the relations between bodies and place, or bodies in space, as active, agentic, and dialogic:

> The subject's relation to space and time is not passive: space is not simply an empty receptacle, independent of its contents; rather, the ways in which space is perceived and represented depend on the kinds of objects positioned 'within' it, and more particularly, the kinds of relation the subject has to those objects. Space makes possible different kinds of relations but in turn is transformed according to the subject's affective and instrumental relations with it. (1995: 92)

The women's peace camp is emblematic of such transformations of space and bodies through a movement, through being activated/activist.

Ecologies of Thinking

The Western anti-nuclear peace camps of the 1980s were organized around theories of collective process, consensus decision-making, non-violent direct action, as they also contributed to questioning and thinking more broadly about ways of coming together to live. Issues like food sources, waste, creativity, forms of resistance, and governance all became implicated as part of an ethics or politics of living. Plumwood's experience also made this link explicit, demonstrating the dialogic relation between bodies, thinking, and place that can be violently transformative. The complexities of thinking through such politics can also be violent, and transformation is not always for the better, but the legacies of such events are surprisingly enduring in public debates and events.

One example of living the politics/politicizing living is the decision to make the camps women-only. As at Greenham, this was an ideological position that critiqued patriarchy. Loreto nun Sister Margaret Hill from Melbourne wrote at the time:

> Women can offer an alternative to the patriarchal system which dominates and characterizes society and has provided the world with its current balance of terror that offers not a future but the possibility of extinction. An all-women's peace protest aims to highlight this. (Hill 1983)

This rhetoric often became entrenched into a universalizing maternalism, associating women with nature and as nature's caretakers akin to a maternal capacity (Roseneil 2000: 45; Murray 2006: 87; Bartlett 2011: 33). The following song lyrics from the camp's song sheet implicitly position women as intrinsically caring for the earth: 'We are women/we are crying/we are singing/for the earth' (Songsheet 1983). This was the very problem that Rich and Grosz were grappling with their challenges to essentialism. Women's Studies academic Bev Thiele was compelled to write to the Women for Survival newsletter in 1984 to critique this caring position as essentialist:

> The idea that women have a unique contribution to make peace because our reproductive capacity somehow naturally makes us peaceful, protective, nurturant, caring and loving … misrepresents women—and it's not just the press, we also misrepresent ourselves if we think that women are naturally nurturant. If nothing else we assume that all women deep down inside support peace and that no woman is involved in the nuclear system. (1984: 14)

Attending to the specificities of women, but also to exemplify the widespread concern of Cold War politics amongst all kinds of women, both

Greenham and Pine Gap were keen to emphasize the social diversity of the women who were present. Megg Kelham, for example, lists the participants at Pine Gap as:

> Hippies, academics, housewives, 'dole bludgers'; Hindu and Christian nuns, new world spiritualists, Quakers and at least one witch; Liberal, Labor, Democrat, Communist Party voters; dispossessed urban Kooris and Indigenous language-speaking occupiers of traditional lands; greenies, feminists, pacifists, anti-nuclear campaigners, seasoned political activists, public purse politicians and 'protest virgins'. (2013: 76–77)

Commentators agree, however, that the media attention hinged on this rhetoric. The scandal of women protesting, in the desert, without their (presumed) husbands and children—or with their children—was taken up by the press as the key story. Making a public place into a homely camp and a woman's place outside was seen to be a scandalous inversion of social protocol.

This turning inside out of usual gender and labour relations also applied to sexual relations. Living closely alongside other women and lesbians, protesters learnt about intimacy between women and formed relationships in a space where loving women was accepted. Greenham's most prolific scholar, sociologist Sasha Roseneil writes of the queer feminisms of Greenham, to account for its overturning of conventional ways and wisdoms at this collective community. She describes the impact of living 'right up against the fences of patriarchal militarism':

> At Greenham, personal life was radically de-privatized—and eating, sleeping, and even toileting were politicized. Food was collectively provisioned, and the politics and ethics of what was eaten were fiercely debated. Conventional family life, and the heterosexuality and monogamy on which it is built, were named and critiqued as women found themselves developing close, sometimes sexual relationships of love and friendship with the other women with whom they were living and protesting. Bodies that sat together around the fire often lay down to sleep together in large communal benders, or just under the stars. Daily ablutions were carried out outside, showers fabricated and strung up in trees, water heated on the fire. Shitpits were dug and moved around, so as to live lightly on the land. (2013: 199)

Roseneil calls this a 'liminal space', a 'women's community', a 'community of protest' (2013: 199) which reshapes the 'architectures of life' (120). It might also be considered a feminist ecology in that everyday life is lived in conscious response to feminist politics, which are sustainable, ethical, and ecologically sensitive, while protesting patriarchal militarism and nuclear war as part of a continuum of violence. Margaret Laware insists that the protest be located 'in a larger context of peace and antimilitarism and connections to life including kindness to animals, veganism, and antiracism' (2004: 35). These kinds of critiques of patriarchal culture, and the production of a women's culture, become possible as a result of embodied activism—particular bodies living intimately in particular protest spaces.

Ideas have the capacity to move us, to create movements like feminism, and to move across time and space to be recomposed as legacies and languages, seeds, and strategies. Epistecology anticipates the potency of particular places, bodies, and ideas as mutually constitutive and transformational: as feminist epistecology.

Kindred Times 2016

In this chapter I have been arguing for the continuing significance of 1980s thinking and the women's peace camps, which lay the groundwork for the reanimated feminist ecologies of the present. Rereading these texts and remembering events remind us of the layers of ideas that have come before. Turning her attention to feminism, activism, and time, Grosz writes that the future is never predictable, but can be understood as 'an unexpected shift [...] which reorients the past and whose reorientation or reanimation reorganizes its present effects' (1995: 258). Those resonances can arise in unexpected ways that continue to form kindred associations and rekindle foundational thinking.

The summer of 2016, when I was writing and protesting at the Roe 8 site, felt like a reanimation of my research from a different time and place. The Roe 8 movement was led by women who organized into affinity groups, ran non-violent direct action training, shared skills for media and police liaison, and brought materials to decorate the mesh fence that was erected to keep us out. We created family-friendly spaces for kids, collected seeds, sang songs, distributed fragments of fabric coded blue, created rosters and networks through encrypted smartphone apps ironically called Telegram. We mounted events for media consumption to bring attention to the wilful destruction of urban bushland and critical wetland systems. Mothers and daughters locked themselves onto the gates where the bulldozers were parked overnight. Neighbours were arrested. A group of professors joined forces to write articles (The Beeliar Group 2017). We learnt new languages of road building, animal trapping, environmental compliance policy, and policing strategies. The dust in my mouth was accompanied by mouthing new words in old languages as we learnt multiple terms for local animals and plants in their common, Latin, and local Noongar names.

> Southern bandicoot, *Isoodon obesulus*, quenda.
> Grass tree, *Xanthorrhoea preissii*, balga.
> Christmas tree, *Nuytsia floribunda*, mudja.

On the radio I heard a programme about environmental grief and the new terms that are being used to express this collective feeling of mourning: solastalgia. Biophobia. Ecoanxiety. Global dread. Ecoparalysis (Earshot 2016). New languages are being invented for new forms of embodied violence, and perhaps new thinking about urban ecologies can be generated as a result.

In trying to make sense of the emotional impact of the summer, of why it is so harrowing to hear giant marri trees fall under a bulldozer, or to witness ancient balgas mulched into a pile of woodchips, I came across Donna Haraway's response to the concept of the Anthropocene (2015). Haraway's famous *A Cyborg Manifesto*, which controversially ended with the slogan 'I'd rather be a cyborg than a goddess' (1985: 101), was written in 1983 and published in 1985 in *Socialist Review*, around the same time that Adrienne Rich was making her 'Notes Toward a Politics of Location'. Like Rich, Haraway's thinking about the connections between cybernetics and organisms began with the military industrial complex, and a kinship with the women of Greenham common (Feigenbaum 2015: 270). In her recent thinking, Haraway challenges the Anthropocene and other namings of epochs in which unprecedented change is being marked geologically in the layers of the earth. She suggests that we live in an age of the Chthulucene which 'entangles myriad temporalities and spatialities and myriad intra-active entities-in-assemblages—including the more-than-human, other-than-human, inhuman, and human-as-humus' (2015: 160). The Anthropocene is a human-centric theory, she argues, that does not give enough credence to the other creatures or unseen worlds with whom we live and depend. We are now a different kind of human to that of the 1980s, one whose body has been found to have only 10 per cent of uniquely human genomes, with the other 90 per cent shared with companion species: bacteria and fungi and other a/biota with whom we coexist, on whom we depend to exist, so that 'To be one is to *become with* many' (Haraway 2008: 4).

In this sense, Haraway's thinking might have something in common with the concerns of Aileen Moreton-Robinson (2013) when she outlines an Australian Indigenous women's standpoint theory. Moreton-Robinson uses Haraway's 'situated knowledges' to critique the way that some eco-feminists subscribe to a 'body/earth split', which discursively constructs 'the "earth" as a single metaphysical concept that is not embodied' (2013: 335). This is further predicated on a plane of privilege that include 'her relations to land as private property and the nation's sovereignty' (335). In contrast, Moreton-Robinson claims that an Indigenous women's standpoint theory is 'not predicated on the separation of ourselves from our countries, human ancestors, creator beings and all living things' (344) but sustained through a 'bloodline to that country through creator and ancestral birth' (335). One basis for this epistemology is an understanding of self. Moreton-Robinson claims that the Western definition of the self as multiple, and in a process of becoming, is essentialist:

> This conception of self, whose humanness is disconnected from the earth, values itself above every other living thing, is a form of strategic essentialism that can silence and dismiss non-Western constructions which do not define the self in the same way. Such silencing is enabled by the power of patriarchal knowledge and its ability to be the definitive measure of what it means to be human and what does and what does not constitute knowledge. (343)

Moreton-Robinson's proposal speaks to the premise of epistemological ecologies, in that place features as a factor in the production of knowledge. Working towards this position as a non-Indigenous theorist has meant Haraway stretches language (and therefore concepts) to expand what is imaginable. Her new mantra to 'make kin, not babies' recognizes our 'biotic and abiotic sym-poietic collaborators, and co-laborers' (2015: 161) with which we cohabit a complex ecology. Her poetics call for feminists of imagination, theory, and action to lead us in the Chthulucene. Moreton-Robinson calls for feminists to acknowledge our location in relation to land and privilege. They both demonstrate Grosz's claim that '[k]nowledges are not purely conceptual nor merely intellectual ... knowledge is an activity; it is a *practice* and not a contemplative reflection. It *does things*' (1995: 37). In this chapter, the activism of the women's anti-nuclear peace camps in the United Kingdom and Australia demonstrates theory in action and active theory which *does things* in places and times. Furthermore these legacies continue in our thinking and our actions, as I found at the Roe 8 protests, as we continue to think through our relations to land and privilege that might just connect us to country, as kin, through epistecologies of feminist thinking.

Bibliography

Badiou, Alain. 2005. *Being and Event*. Trans. O. Feltham. London: Continuum.

Bal, Mieke. 2002. *Travelling Concepts in the Humanities: A Rough Guide*. Toronto: University of Toronto Press.

Bartlett, Alison. 2011. Feminist Protest and Maternity at Pine Gap Women's Peace Camp, Australia 1983. *Women's Studies International Forum* 34: 31–18.

———. 2013. Feminist Protest in the Desert: Researching the 1983 Pine Gap Women's Peace Camp. *Gender, Place and Culture* 20 (7): 914–926.

———. 2016. Remembering Pine Gap: A Feminist Activist Exhibition. *Continuum: Journal of Media and Cultural Studies* 30 (3): 307–315.

Bolt, Barbara. 2000. Shedding Light for the Matter. *Hypatia* 15 (2): 202–216.

Code, Lorraine. 2008. Thinking About *Ecological Thinking*. *Hypatia* 23 (1): 187–203.

Cresswell, Tim. 1996. *In Place/Out of Place: Geography, Ideology, and Transgression*. Minneapolis: University of Minnesota Press.

Earshot. 2016. Climate of Emotion: Despair. *ABC Radio National*. Broadcast 31 October. http://www.abc.net.au/radionational/programs/earshot/climate-of-emotion:-despair/7880378

Feigenbaum, Anna. 2015. From Cyborg Feminism to Drone Feminism: Remembering Women's Anti-Nuclear Activisms. *Feminist Theory* 15 (3): 265–288.

Grosz, Elizabeth. 1987. Notes Towards a Corporeal Feminism. *Australian Feminist Studies* 2 (5): 1–16.

———. 1994. *Volatile Bodies: Towards a Corporeal Feminism*. Sydney: Allen & Unwin.

———. 1995. *Space, Time and Perversion: The Politics of Bodies*. St. Leonards: Allen & Unwin.

Haraway, Donna. 1985. Cyborg Manifesto: Science Technology and Socialist Feminism in the 1980s. *Socialist Review* 15 (2): 65–107.

———. 1988. Situated Knowledges: The Science Question in Feminism and the Privilege of Partial Perspective. *Feminist Studies* 14 (3): 575–599.

———. 2008. *When Species Meet*. Minneapolis: University of Minnesota Press.

———. 2015. Anthropocene, Capitalocene, Plantationocene, Chthulucene: Making Kin. *Environmental Humanities* 6: 159–165.

Hill, Margaret. 1983. Untitled. Canberra Women for Survival Papers, Box 0023 File 7, Jessie Street National Women's Library Archive.

Kelham, Megg. 2013. War and Peace: A Case of Global Need, National Unity, and Local Dissent? A Closer Look at Australia's Greenham Common. *Lillith: A Feminist History Journal* 19: 76–90.

Laware, Margaret L. 2004. Circling the Missiles and Staining them Red: Feminist Rhetorical Invention and Strategies of Resistance at the Women's Peace Camp at Greenham Common. *NWSA Journal* 16 (3): 18–41.

Moreton-Robinson, Aileen. 2013. Towards an Australian Indigenous Women's Standpoint Theory: A Methodological Tool. *Australian Feminist Studies* 28 (78): 331–347.

Murray, Suellen. 2006. 'Make Pies Not War': Protests by the Women's Peace Movement of the Mid 1980s. *Australian Historical Studies* 37 (127): 81–94.

Plumwood, Val. 1993. *Feminism and the Mastery of Nature*. London: Routledge.

———. 2012. *The Eye of the Crocodile*. Ed. L. Shannon. Canberra: Australian National University E Press.

Rich, Adrienne. 1985. Notes Toward a Politics of Location. In *Women, Feminist Identity, and Society in the 1980's: Selected Papers*, ed. M. Díaz-Diocaretz and I. M. Zavala, 7–22. Amsterdam: John Benjamins.

Roseneil, Sasha. 2000. *Common Women, Uncommon Practices: The Queer Feminism of Greenham*. London: Cassell.

———. 2013. Architectures of Resistance and Transformation. In *Potential Architecture*, ed. L. Orta and J. Orta, 117–121. Bologna, Italy: Damiani.

Shaw, Tony. 2007. *Hollywood's Cold War*. Edinburgh: Edinburgh University Press.

Somerville, Margaret. 1999. *Body/Landscape Journals*. North Melbourne: Spinifex.

Songsheet. 1983. Canberra Women for Survival Papers, Box 0023 File 15. Jesse Street National Women's Library Archive.

Sutter, Paul S. 2014. The Tropics: A Brief History of an Environmental Imaginary. In *The Oxford Handbook of Environmental History*, ed. A. C. Isenberg, 178–205. Oxford: Oxford University Press.

The Beeliar Group. 2017. Accessed May 31, 2017. https://thebeeliargroup.com/

Thiele, Bev. 1984. Women, Nature and the Peace Movement. *Social Alternatives* 4 (3): 13–16.

Alison Bartlett is Associate Professor of Gender Studies at the University of Western Australia. Her research on the 1980s women's peace camps in Australia is widely published. She is the author of *Jamming the Machinery: Contemporary Australian Women's Writing* (1998) and *Breastwork: Rethinking Breastfeeding* (2005), and has edited books on the public sphere, postgraduate pedagogy, maternal ethics, feminist material culture, and most recently journal issues on social memory (*Continuum* 2016) and feminist museology (*Journal of Australian Studies* 2016).

10

Performing Ghosts, Emotion and Sensory Environments

Peta Tait

Art work by women artists that aligns with ecofeminist values seems even more potent in the twenty-first century because it grapples with gender identity, feminized views of nature and an impending environmental crisis. The feminist analysis of identity politics and power relations is relevant to social responses to scientific predictions of climate change and habitat decline for human and nonhuman species. This chapter explores the ways in which contemporary performance and art by women artists contributes productively to increased public responsiveness to environmental concerns. The performances and photographs by Australian artists Jill Orr and r e a (Gamilaraay) discussed in this chapter encapsulate the feminist rejection of the ensnared categorization of 'woman' as 'nature' while they reflect thinking about embodied identity and social inclusion.[1] By presenting a solitary figure in a natural world, these site-specific performances and photographs reflect some of the complex postcolonial and posthuman dimensions of recent feminist politics as well as the

P. Tait (✉)
La Trobe University, Melbourne, VIC, Australia

L. Stevens et al. (eds.), *Feminist Ecologies*,
https://doi.org/10.1007/978-3-319-64385-4_10

increased focus on environmental concerns in the arts. This chapter asks: how can looking at performance and photographic art that reflects feminist and ecological values contribute to a political understanding of the human and the nonhuman? It argues that body-based performance that unsettles and emotionally disturbs can draw the viewer's attention to how the environment is always phenomenologically perceived through the body's sensory and emotional processes. In this way, an eco-phenomenological approach builds on earlier ideas of ecofeminism.

Orr's *The Promised Land*, first created in 2011, depicted a timeless and ambiguously gendered figure in a boat that referenced a long cultural history of seeking new land as it hinted at the political legacies of an ensuing occupation. In *PolesApart*, first exhibited in 2009 by Aboriginal multimedia artist r e a, a woman in a long black dress was pursued among fire-blackened gum trees. The larger meaning of this figure being hunted resonated with Australia's colonial occupation, loss of land and species habitats. In these two art works, ghost-like humans appeared to haunt darkened and fire-affected nonhuman spaces.

Shared Ecologies

As Freya Mathews argues about ecofeminist philosophy, irrespective of the value humanity might accord the natural world as a whole, ecofeminism recognizes difference within it and assumes that nature's intrinsic value is derived from the multiplicity of living entities (see Chapter 3). She contrasts ecofeminist ideas of an interconnected 'family of life' with the philosophy of deep ecology that upholds an understanding of the whole natural environment as it is experienced by the human self. If the latter is potentially more androcentric, the shared natural world envisaged by ecofeminism points directly to the possibility of unravelling human power relations in the dominance of habitat and other species. The problem of how to reach beyond or get outside the dominance of the human self was addressed by earlier ecofeminism in, for example, the rejection of dualistic categories of thought in Mathews' work and the rejection of patriarchal sociopolitical structures evident in Ariel Salleh's approach (see Chap. 1). The feminist position presented in this chapter contends that the human body-self continues to be a material problem within shared ecologies when it remains oblivious to the dominance of its sensory and emotional processes—functions also shared with nonhuman animal species.

The movement for animal rights was also developing contemporaneously with early feminism and it expressly drew on notions of sexism to explain speciesism, so the distinction between environmental concerns and nonhuman animal rights may be less pronounced in ecofeminist approaches (Plumwood 1993). As ecofeminism points out, there is blinkered social appreciation of how the environment must be shared with other species; countless animal species are now threatened. In her concise history of Australian environmental philosophy—of an ecophilosophy—and in both the European and Aboriginal traditions, Mathews (2014) argues for their integration in order to recognize the diverse sentience of the natural world. Systems of thought and communication need to reinforce the way in which environmental space is shared bodily within the context of biodiversity and species survival. Performance and art offer important means of communicating about shared ecologies.

Una Chaudhuri (2017) suggests, however, that political performance about nonhuman animals does not need to present them to reveal shared ecologies. She points out that the ethical problems of unnaturally staging most animals override the political efficacy of increased visibility in human worlds through performance. As John Berger (2009) recognized about art, it is crucial to challenge the fundamental human assumption that animal species and their environments should be looked at by humans. Berger's earlier groundbreaking work considered how female bodies were presented and looked at in visual art. Therefore an ecofeminist argument in the twenty-first century might interrogate gendered and speciesist patterns of dominance through looking in relation to the nonhuman, and, further, it might do so by exploring embodied processes of seeing and perceiving the environment and other species.

In *Nature*, Maurice Merleau-Ponty points out that the 'faculties of knowing', which produced humanist ideas starting with Descartes, position human beings in opposition to 'the phenomena of nature' (2004: 25). The history of rationalist thought did not situate humans in nature.

By contrast, in the 1950s Merleau-Ponty argues that lived experience and nature can be more fully understood by exploring what he calls the human 'corporal schema' and a 'theory of the flesh' (2004: 209). Merleau-Ponty writes: 'The body is not only a thing, but also a relation to an *Umwelt*; this is already true of an animal body' (2004: 209). This kind of relational bodily awareness necessitates sensitivity to other bodies within the environment while it mitigates against notions of transcendence and oneness. Merleau-Ponty outlines how this might be achieved by encounters with nonhuman animals and through art. As Louise Westling argues, Merleau-Ponty's phenomenology avoids what Val Plumwood criticizes as deep ecology's 'erasure of distinctions' (2014: 43). Bodily self-awareness necessitates sensitivity to other nonhuman entities and bodies within the environment.

I apply Merleau-Ponty's theories of embodied phenomenology to thinking through the ways in which spectators watch live performance and view art. Such embodied phenomenologies can make an observer aware of the physiological effects of bodily looking and emotionally feeling in response to a live performance or an art object. In the same way, an encounter with a different species body might cause the automatic phenomenological sensory engagement in the lived world to stall momentarily and make an observer self-aware that he or she is enfolding and unfolding the flesh of the world; that is, engaged in a process of 'fleshing' the surroundings (Tait 2015). This type of self-awareness about looking can also arise in a heteronormative context when an observer cannot pre-cognitively or automatically recognize the gender identity of a human figure and is forced to pause and think within the process.

Bodily engagement within the world might be fundamental to human life as both Merleau-Ponty's phenomenological philosophy and feminist theory about identity difference argue, but such all-powerful emotionally affective, visceral dimensions continue to be minimized in Western culture which still privileges the rational and empirical proof. Further, the physiology of the senses and emotions may be culturally conditioned over time by bodily experiences of gender, race, sexuality and species identities. These in turn impact on our relations within a given environment. The denial of these bodily processes makes it possible to ignore the impact of the material needs of aggregate human bodies.

This chapter argues that an artistic disruption of the sensory patterns of seeing can contribute to the ecofeminist effort to make humans aware of how they are bodily and materially situated in shared environments because this disruption is compounded by the ways performance and art arouse affect and emotional responses.[2] In particular, the stimulation of bodily responses to performance and art about the environment can draw attention to phenomenological processes of bodily seeing and perceiving in other circumstances. Performance and art suggests emotionally felt reactions may influence the capacity to bodily see and perceive an environment.

Queer Haunting

Orr's body-based performance and visual arts practice spans five decades, and while I have previously considered a particular set of her performances using theory about body phenomenology (Tait 2015), Anne Marsh has astutely analysed her oeuvre from feminist theory's multiple perspectives. Most recently, Marsh explains that Orr's art 'engages with ecological and social issues as well as rites of passage and gender' (2014: 99). *The Promised Land* consisted of a silent live gallery performance by Orr sitting, standing, rowing and lying within a plywood frame of a rowing boat, and a photographic installation. Orr wore a full-length, full-sleeved coat and head covering suggesting non-specific historical clothing and performed surrounded by photographic images of the performance in site-specific environments. The live performance was presented in Melbourne in 2011 and in Venice in 2012.[3] One photograph shows Orr with the oar used in the live performance, another shows her waving a large red flag reminiscent of a propaganda image from the Russian Revolution, and in a further two photographs, she is carrying a sack on her back at a riverbank and at a boat dock.

A ghostly effect was created by the transparent pale wood skeleton of the boat, and Orr's white costume made her seem both timeless and gender ambiguous. It was unclear if the figure standing in the boat was someone working or travelling, leaving or arriving, fleeing or finding sanctuary. It might have been someone in authority, even a religious figure: a priest or a Sufi mystic. The props, however, added an extra dimension as the sack suggested a labourer or a miller and the flag suggested a historical revolutionary. In the photographs, the white figure in the foreground contrasted with the dark tones of the background vistas captured at twilight. They were familiar but not realist photographs—evoking a sense of the uncanny. The boatman, for example, might suggest the Western cultural narrative of crossing the river Styx into the afterlife (Fig. 10.1).

The significance of the natural world in Orr's oeuvre is unmistakable. In performances since the 1970s, Orr's work has used visual juxtaposition to draw attention to the biological human body within the natural environment, and in the 2000s she controversially worked with dead animal carcasses in gallery spaces (Tait 2015). At times, the placement of the naked female body buried in the earth or covered in whitened ash or bodily suspended above sand depicted violent intrusions and relations within nature. More recently Orr's beautiful filmed and photographed performance shifted to presenting figures in Victorian female dress, and

Fig. 10.1 Jill Orr, *The Promised Land—moving*. Photographer: Christina Simons for Jill Orr (©)

in gender- and species-ambiguous costuming within desert-like landscapes that include the waste from gypsum mines.[4]

In her exploration of ideas of toxicity, bodies and matter, Stacy Alaimo cites Lynda Birke writing against '[t]he ghost of biology' that haunts femaleness indicative of the feminist effort to 'disentangle *woman* from *nature*' and reject gendered binaries (2010: 5, italics in original). Alaimo outlines the need for an expanded theoretical response in the context of environmental pollution and industrial worker ill health. She presents feminist ideas of 'trans-corporeality' as bringing together theories of the biological body and the environment and as a means to reinsert nature into feminist theory. Alaimo draws on the scientific understanding that biological nature is not passive or static and cites Myra Hird explaining that at a cellular level even the human is composed of a queer nature that is active and productive (2010: 5). Cellular life that exists on and across bodies does not follow gendered protocols. If gender identity itself became a ghostly apparition in *The Promised Land*, the action of rowing in the live performance initially made the figure seem more male than female. But as Orr's action changed to curling up as if fearful, pulling up a cover to hide, and then lying out as if dead, the emotional impressions undermined the initial suggestion of masculine fortitude; the unfolding energetic action reinforced impressions of an ambiguous, unstable queer identity.

But the uncertainty about the identity of the queer ghost-like figure in *The Promised Land* seeped into the surroundings. The darkened waterscapes had an unsettling, even puzzling, effect. As Sara Ahmed (2010) explains, the instabilities of queer human identity traditionally flowed towards disturbing emotions such as unhappiness and disgust in a 'queer phenomenology' because of negative social reactions. Yet such emotive reactions to the instability of the nonhuman can draw focused attention. The visual eeriness of the photographs aroused suspicion. The imagery suggested harm possibly an underlying peril at a cellular level that lends itself to an interpretation with new materialist ideas of energetic flows even though it seemed fraught (Bennett 2010; Alaimo 2010). This was not only disgust for decay but seemed dangerous as if the water might be toxic. As Mathews describes, there is a long-standing repression of the darker aspects of cellular life in death, decay and destruction in discourse about the environment (Chapter 3). If emotional rejection of cellular decay is typified through responses of disgust and squeamishness, these have become compounded with the knowledge that pollution and contamination often cannot be seen.

While the title, *The Promised Land*, alludes to Judeo-Christian aspiration and the historical idea of a utopian world or afterlife, Orr's troubling imagery was directly relevant to major political events occurring at the time of the performances. In Australia, the performance and photographs suggested environments haunted by the history of colonial culture which took the land of the Aboriginal inhabitants, and damaged the ecosystems with unsuitable European farming methods and introduced species. In both the Australian and the Venice performances, it was a reminder of recent events in which refugees escaping war-torn countries struggle to reach safe lands by boat—especially parts of Italy. The darkened environments in the photographs of *The Promised Land* seemed beautiful but menacing.

Orr's work lends itself to numerous interpretations. The human figure in the boat was marooned at the edge of land as if threatened or unable to venture further, inviting the viewer to consider the precariousness of human existence in the encounter with the nonhuman. The significance of the landscape became ambiguous like the queered human figure as it evoked emotional unease in a twenty-first-century viewer conversant with climate change and rising oceans and failing river flows. As the solitary figure evoked historical human presence in past environments, it seemed to warn about a future with only human ghosts. The juxtaposition of the queer ghost and the nonhuman evoked foreboding as *The Promised Land* traversed time and water.

Felt Natures

PolesApart is a complex, open-ended gallery installation by the Australian artist, r e a, consisting of a six-minute filmed performance without sound, a set of photographs and a written narrative.[5] At the beginning of the performance, the viewer gradually became aware of a moving figure in a natural landscape of trees with fire-effected bark, and it directly presented the plight of this woman as it indirectly evoked other species and climate change. The surprise ending of the performance cleverly and playfully interwove political resistance to colonial and national identity with ideas of artistic heritage to challenge traditional ways of looking at the human, the nonhuman and gender identity.

Multimedia artist, r e a , is from the Gamilaraay people of northern New South Wales (NSW). When the *PolesApart* installation was first presented in 2009, it consisted of a filmed performance and photographs, accompanied by a written narrative about the artist's family.[6] Peggy Phelan writes that r e a 's art depicts multiplicity 'in a context of radical political difference' (2001: 19). Her highly accomplished oeuvre is a sustained exploration of Aboriginal, queer and self-identity. Her twenty-first-century art works deliver heightened sensory experience through explorations of the human in a natural environment.

The silent filmed performance began with imagery of a sloping hillside covered in Australian eucalyptus trees with sparse undergrowth. It gradually showed that some of the trunks were burnt black, as if a bush fire had burnt through in recent years, a familiar landscape in Australia. A woman in a full-length, nineteenth-century black dress appeared among the trees. As she lifted the dress to run, it gradually became evident that she was stopping to hide behind the trees, as if pursued by someone. She repeatedly looked back; she looked in the direction of the viewer. The human figure running in this landscape conveyed the significance of being hunted especially when she tripped and fell. The environment seemed hostile because of her running action. In an eco-phenomenological response, anxiety about the fate of the woman might have spilled over into anxiety about her surroundings. The performer's movement directed the viewer's attention to the trees or the ground; when she fell the screen image flickered in mimicry of faltering liveness.

Some viewers of the filmed performance might have built a narrative in which the threatened woman escapes and flees into a forest. Certainly the Australian bush was not considered a safe environment to colonials, but the woman seemed to be afraid of someone. But who was following her? Her black dress resembled mourning attire; it matched the burnt black tree trunks and evoked loss and death. To an Australian viewer, the action points to the white settler persecution and abduction of Aboriginal

Australian people. The woman's close-cropped hair defied the femininity of the dress, although it might have suggested the brutal institutional practices endured by Aboriginal people. But it was the combination of the running and the hiding action of the female figure that brought to mind dispossession and genocide, as well as the sexual ownership of Indigenous women (Conor 2016). At the same time it conveyed impressions of the inherited cultural practices of ecological sustainability and ideas of custodianship, land utilization developed over millennia in specific environments in ways that preserved them (Pascoe 2014). Land management that was suitable for fragile Australian ecosystems abruptly ended with colonization.

There are multiple possible meanings for this performance. Significantly, r e a sees regeneration in this bush of blackened trunks; she perceives a type of love and hopefulness in the way trees come back to life.[7] As a white migrant to Australia, I see it differently and sometimes fear this bush that can burn fiercely, while r e a holds an opposite emotional perception. Her art expands the emotional possibilities of how to perceive environments.

In the 2009 installation, r e a 's performance of running in the bushland could be also interpreted as a re-enactment of a personal family narrative, although this art work was definitely not storytelling and the installation only provided an indirect link to family experience. Instead the performance was emotionally evocative. In 2009, the written material described the lives of r e a 's grandmother, Ruby, and her great-aunt Sophie. In 1916 they were removed from their homes at 5 years of age and taken to the Cootamundra Girl's Home where they were trained to be servants and sent into service. Sophie decided to escape this servitude and did succeed in returning home, whereas Ruby went on to work as a servant, including working for a time for the Australian opera singer, Nellie Melba. It was clear that r e a 's family background involved the stolen generation where children were forcibly removed from their families by the government and therefore from their 'country' as it is called by Australian Aboriginals (Nicholls 2009; Kabaila 2012). While the personal narrative is implicit in this performance, the larger historical and political meanings of the art work resonate with clarity and profound significance. The historical mourning dress suggested that the past haunts the present; it seemed that grief and suffering were in pursuit of the woman.

A spectator's body-based phenomenological response might have started out by appreciating the pleasant natural vista of a nondescript landscape before he or she began to notice that the tree trunks were burnt. As the woman caught the viewer's attention, her movement might have engaged with the viewer's body memory of sensations of running, including those of catching the breath. But the performance positioned the viewer in a process of looking that converged with chasing. The sensible movement gained the viewer's attention and then seemed to make the viewer complicit with what was happening to the woman. There might have been a growing sense of unease. At some point, he or she might have become aware of attributing emotions of nervousness, even terror, to the woman because she seemed to be escaping something and hiding. Curiosity about the figure might have turned into sympathy and concern about whether she would be caught.

In her exploration of a technologically constructed art work that allows a spectator to immersively enter a virtual world from another's visual perspective, Sigrid Merx extends the possibility of 'doing phenomenology' to 'doing empathy' (2015: 217, 218). She draws on Matthew Ratcliffe's ideas of empathy as 'an act of imagination in which we project our own experiences on someone else' that leads to a personal first-person 'radical empathy', which is a 'distinctive kind of empathy that requires letting go of our habitual understanding of the world around us' (2015: 215, 216, 217). While Merx's viewer was put into a virtual world, it is possible that *PolesApart* evoked a comparable sensory emotional effect in the spectator watching the running action of the woman.

In pointing to what Mary Midgley's and Plumwood's ecofeminism criticizes in liberal political thinking that downplays gender, Rosi Braidotti discerns the possibility of a '*zoe*-egalitarian turn' which includes what she terms the posthuman (2013: 71, 77, italics in original). She explains that this socially just and enlarged world will require empathy if it is to be realized. As it pointed to a relationship of human and landscape, the performance by r e a lends itself to an expansion of empathy out towards the environment, an effect created by the woman in black moving through and gently touching trees blackened by fire for an extended time. The running evoked all dispossessed inhabitants of the past, including nonhuman animals. It prompted the question: have animals disappeared from this environment given that they have limited chance of surviving uncontrolled big fires?[8] As Mathews points out, human 'interconnectedness' with all life is central to Aboriginal philosophy (2014: 581), and therefore the removal of animals would become a grievable loss to humans.

At the end of the performance, the woman stopped as if caught. But the emotional narrative of concern for the woman did not unfold in a predictable way; it was completely overturned. The woman's black dress was suddenly squirted with white paint, then blue paint, then red paint—like a canvas. The woman laughed but the action was serious. The white substance over the black cloth became the unavoidable threat and impact of whiteness for Indigenous peoples, with the red substance evoking blood spilt and absorbed, the blue reinforcing an ominous quality. The woman's identity collided with the colours of the Australian flag—the symbolic object of European colonization and Australian nationalism—as it merged with covert references to the history of Australian art about the environment. For example, the Heidelberg school of painters including Frederick McCubbin who famously depicted masculine and iconic white settler figures in a comparable vista of Australian bush (Astbury 1985).

A viewer's bodily response to the ending, however, might have been a startle or a jolt, or even visual avoidance of the squeamish effect of the squirted paint. The ending challenged a viewer to become self-aware of his or her bodily responses as it cut through expectant assumptions and emotional processes. Such bodily affect was dispersed into surprise and puzzlement, possibly annoyance or amusement, reminding a viewer that the political efficacy of art involves sensation and emotional feeling. As such, the performance overturned the emotional psychologies of the expected, predictable resolution at the end. It reconfigured the historical colonial collapsing together of the categories of Indigenous woman and nonhuman animal in nineteenth-century iconography by presenting the Aboriginal woman as the creative artist who is in control. The performance asked the viewer to consider bodily patterns in 'a double movement' (Toadvine 2009: 78), taking place between the viewer and the performer, the human and the nonhuman, the visible and the invisible in the environment. Through the gender and race implications, the eco-phenomenological process also becomes ecofeminist because the politics are an inseparable part.

In performing emotional resistance to the force of history, r e a 's art offers a way to confront the limitations of colonial abuse of both Aboriginal women and country. *PolesApart* points to the perceptual knowledge of Indigenous peoples about the natural world and potentially offers a way forward in the Anthropocene age by promising a different emotional relationship to the environment leading to alternative ecological practices and action for climate change. Among his proposals for how to support climate change activism, George Marshall writes 'climate change generates strong feelings that can, unless recognized, lead us to disavowal and outright denial. We need to recognize people's feelings of grief and anxiety, and acknowledge and provide space for contradiction, ambivalence, loss, and mourning' (2014: 238). The failure to recognize the centrality of human emotions in attitudes to the environment is damaging it; the full spectrum of emotions must be accommodated at this time.

I suggest that, as r e a 's performance invites phenomenological sympathy even empathy for the running woman, such emotional responses spill over into the surroundings as they eventually challenge pre-existing patterns of seeing. The surprise ending made a viewer aware of emotional assumptions about what was being looked at, as these were dispelled and dynamically replaced in the evocation of unexpected feeling. The performance directed attention to layers of human emotion in the perception of environments. This is activism through art.

Emotionally Feeling to See

Orr's and r e a 's performances presented the viewer with recognizable possibilities as narrative threads multiplied or reversed so that the landscape became inflected with this indeterminacy. These silent performances invited a viewer to look at a human figure in relation to the natural environment in ways that highlighted oscillations between sensory responses, emotional feeling and cognitive knowing. There was no implicit or explicit information and feeling from either words or music. In a phenomenological framework, the energetic instabilities of a figure rowing or a figure running directly linked with the instability of a viewer's energetic body feelings. Encounters that dispel predictable patterns, including emotional ones such as those provided by these two performance works, increase the likelihood of a viewer's self-awareness of affect, sensation and emotional feeling. They might lead to a realization of 'I don't know how I feel emotionally' after viewing the performance, or 'the way in which I emotionally feel is tenuous'. The performances encouraged the questioning of emotional responses in relation to sensory worlds and the environment.

Ted Toadvine explains that Merleau-Ponty interpreted how we exist within embodied relational patterns in an environment as preceding cognitive function and arising with felt sensory experience (2009: 131). Perception is expressive but reversible. I agree that this embodied orientation to and within environments means that humans are attuned to living worlds and therefore patterns of energetic movement large and small, including at a cellular level. Habitual patterns of perception developed in relation to motility and bodily encounters with others create an automatic expectation of sensing other living nonhuman species in the environment. I suggest that what is static in the environment and even species absence also has the potential to disturb the body's sensory patterns at a precognitive level of perception and at a cellular level and infect emotional responses.

An emotional feeling of discomfort and even disturbance arises when a preset sensory expectation is thwarted. The body-self's sensory perception becomes troubled and wary when patterns are disrupted and observed movement becomes unpredictable and overturned. Importantly, as performance and art do the work of orientating the sensory focus, sensation and emotional feeling facilitate awareness of the way in which the body perceives and sustains such experiences. If emotionally felt responses towards the nonhuman can be potent and potentially provide a galvanizing force for protective action, performance and art allow for contradictions within such experience to be situated within the larger political effort and assisted by ecofeminist thinking. These art works highlight the effect of troubling emotion so that it can be recognized rather than repressed. In the examples of *The Promised Land* and *PolesApart*, what is emotionally felt becomes interchangeable with what is bodily seen within the environment.

Notes

1. The artist r e a writes her name in all lowercase with a space between each letter. This is the artists' professional name used to make reference to her practice, which examines the colonization and categorization of the (colonizing) English language.
2. For a discussion of the terms 'emotions', 'emotional feelings', 'sensations' and 'affect', see Tait (2016); for a recent discussion of art and performance and nature and ecology as concepts of culture and some complications for ideas of the environment, see Lavery and Finburgh (2015).
3. I viewed the photographs at Jenny Port Gallery in 2012 but filmed the live performance at the Venice International Performance Art Week, Palazzo Bembo, Venice, 8–15 December 2012.
4. Orr in conversation with the author, 22 January 2016. Gypsum is used for plasterboard.
5. I worked with r e a in contemporary performance group, The Party Line, in the 1990s, and she continues to work with performance director, Gail Kelly, for some of her art works and for *PolesApart*.
6. Even though the whole art work is the film, photographs and autobiographical commentary, only the filmed performance was presented in the Art Gallery of NSW in 2015.
7. Conversation with the author, 30 April 2015.
8. One of the most controversial environmental issues in Australia in the twenty-first century is the extent to which controlled burning is able to prevent the new type of catastrophic bush fire that killed 173 people on Black Saturday, on 9 February 2009. These catastrophic fires are being attributed to climate change. It has only recently been understood more widely in Australia that Aboriginal people managed their environment over thousands of years through the careful use of fire.

Bibliography

Ahmed, Sara. 2010. *The Promise of Happiness*. Durham: Duke University Press.

Alaimo, Stacy. 2010. *Bodily Natures: Science, Environment and the Material Self*. Bloomington: Indiana University Press.

Armstrong, Susan, and Richard Botzler, eds. 2008. *The Animal Ethics Reader*. London: Routledge.

Astbury, Leigh. 1985. *City Bushman: The Heidelberg School and the Rural Mythology*. Melbourne: Oxford University Press.
Baker, Steve. 2000. *The Postmodern Animal*. London: Reaktion Books.
Bennett, Jane. 2010. *Vibrant Matter*. Durham: Duke University Press.
Berger, John. 2009. *Why Look at Animals?* London: Penguin.
Braidotti, Rosi. 2013. *The Posthuman*. Hoboken: Wiley.
Chaudhuri, Una. 2017. *The Stage Lives of Animals*. London: Routledge.
Conor, Liz. 2016. *Skin Deep: Settler Impressions of Aboriginal Women*. Perth: The University of Western Australia Publishing.
Kabaila, Peter. 2012. *Home Girls: Cootamundra Aboriginal Home Girls Tell Their Stories*. Canberra: Canprint Publishing.
Lavery, Carl, and Clare Finburgh. 2015. Introduction: Greening the Absurd. In *Rethinking the Theatre of the Absurd: Ecology, the Environment, and the Greening of the Modern Stage*, ed. C. Lavery and C. Finburgh, 1–58. London: Bloomsbury Publishing, ebook accessed 1 September 2016.
Marsh, Anne. 2014. *Performance, Ritual, Document*. Melbourne: Macmillan.
Marshall, George. 2014. *Don't Even Think About It: Why Our Brains Are Wired to Ignore Climate Change*. New York: Bloomsbury.
Mathews, Freya. 2014. Environmental Philosophy. In *History of Philosophy in Australia and New Zealand*, ed. G. Oppy and N. Trakakis. Dordrecht, NL: Springer.
Merleau-Ponty, Maurice. 2004. *Nature*. Ed. D. Séglard and Trans. R. Vallier. Evanston, IL: Northwestern University Press.
Merx, Sigrid. 2015. Doing Phenomenology: The Empathetic Implications of CREW's Head-Swap Technology in 'W' (*Double U*). In *Performance and Phenomenology*, ed. M. Bleeker, J. F. Sherman, and E. Nedelkopoulou, 204–221. London: Routledge.
Nicholls, Christine. 2009. *Catalogue Essay. PolesApart Catalogue*. Sydney: Breenspace.
Orozco, Lourdes. 2013. *Theatre and Animals*. Basingstoke: Palgrave Macmillan.
Pascoe, Bruce. 2014. *Dark Emu*. Broome: Magabala Books.
Phelan, Peggy. 2001. Survey. In *Art and Feminism*, ed. H. Reckitt and P. Phelan, 12–49. London: Phaidon.
Plumwood, Val. 1993. *Feminism and the Mastery of Nature*. London: Routledge.
Sunstein, Cass R., and Martha C. Nussbaum, eds. 2004. *Animal Rights: Current Debates and New Directions*. New York: Oxford University Press.
Tait, Peta. 2015. Fleshing Dead Animals: Sensory Body Phenomenology in Performance. In *Performance and Phenomenology*, ed. M. Bleeker, J. F. Sherman, and E. Nedelkopoulou, 111–120. London: Routledge.

———. 2016. Introduction: Analysing Emotion and Theorizing Affect. *Humanities* 5 (3): 70.
Toadvine, Ted. 2009. *Merleau-Ponty's Philosophy of Nature*. Evanston, IL: Northwestern University Press.
Westling, Louise. 2014. *The Logos of the Living World*. New York: Fordham University Press.

Peta Tait is a professor at La Trobe University, a Fellow of the Australian Academy of the Humanities and a visiting Professorial Fellow at the University of Wollongong. She has written 60 scholarly articles and chapters, 10 plays and performance texts, and her recent books include: *Fighting Nature: Travelling Menageries, Animal Acts and War Shows* (Sydney University Press 2016); the coedited *The Routledge Circus Studies Reader* (Routledge 2016); *Wild and Dangerous Performances* (Palgrave Macmillan 2012); *Circus Bodies* (Routledge 2005); and *Performing Emotions* (Ashgate 2002). She is editor of 'Great Stage Director Series volume one: Stanislavsky, Antoine and St Denis' and author of 'Emotion' (Bloomsbury Methuen under contract). Her most recent play, *Eleanor and Mary Alice*, about Eleanor Roosevelt meeting with Mary Alice Evatt and human rights, art and war, was restaged in Perth in 2016.

11

You Are on Indigenous Land: Ecofeminism, Indigenous Peoples and Land Justice

Ambelin Kwaymullina

You are on Indigenous land, swimming in Indigenous waters and looking up at Indigenous sky. All those who came to the place now known as Australia post colonization are comparative newcomers to a living land that is formed and informed by the cultures of Aboriginal and Torres Strait Islander nations. And all those who came after inherited the benefits of the taking of the land from those who were here before. How, then, can non-Indigenous ecofeminists ethically advocate for justice in relation to women and the environment when they occupy the fraught position of being continuing beneficiaries of the dispossession of Indigenous women from our homelands? My suggestion is that non-Indigenous scholars must respect Indigenous sovereignty and meaningfully enact this respect, including through the layered process of listening to the voices of Indigenous women.

My thoughts on this matter are grounded in my position as an Aboriginal woman of the Palyku people, but I do not of course speak for all Indigenous women. We are many individuals who belong to

A. Kwaymullina (✉)
University of Western Australia, Perth, WA, Australia

L. Stevens et al. (eds.), *Feminist Ecologies*,
https://doi.org/10.1007/978-3-319-64385-4_11

many nations, and the cultures of Aboriginal and Torres Strait peoples, like those of all Indigenous nations of the earth, are highly diverse. My perspective is particularly shaped by my individual knowledge and experience, the culture and history of the Palyku people from whom I come and the collective inheritance of my ancestors. My voice is part of a continuum of Indigenous voices, and any wisdom I have did not begin—and will not end—with me. Any mistakes are my own.

A Note on Terminology

At an international level, 'Indigenous peoples' refers to peoples who were the inhabitants of a territory when others came there; who were dispossessed; and who continue to maintain distinct cultures in homelands that are now occupied and controlled by others (United Nations Permanent Forum on Indigenous Issues 2010: 4–7). The notion of 'Indigenous' is therefore a category created by colonialism, and the term obscures the vast diversity of the Indigenous peoples of the earth, suggesting a single homogenous culture which does not exist. In order to acknowledge the diversity of Indigenous peoples, any Indigenous person quoted or referred to in this chapter is identified by the specific Indigenous nation or nation(s) from which they come the first time that their work is referenced, provided their specific affiliation was able to be ascertained. If the specific affiliation could not be ascertained or if the person uses a more general term to describe themselves, then the appropriate general term (e.g. Aboriginal, Torres Strait Islander, Indigenous) is used.

Any reference to men or women includes anyone who identifies as male or female, and in this respect, I acknowledge the many struggles of trans and gender diverse Indigenous people who must daily contend with multiple intersecting forms of oppression. Finally, when referring to individuals, their preferred personal pronoun is used.

Indigenous Peoples, Indigenous Homelands and Colonialism

For thousands of years, the many diverse environments of Australia were sustained by the cultures of Aboriginal and Torres Strait Islander nations. Integral to these cultures are the law- and life-ways of Indigenous women. What is sometimes called Women's Business is separate to Men's Business across Indigenous nations but not inferior to it; as Murri Elder Lilla Watson writes: 'from our experience, we can share with the [Western] women's movement a vision of liberated women working on their own terms of reference' (2004: 7). Indigenous women and men alike are charged with caring for the homelands that Australian Aboriginal nations name as 'Countries'. Generations of women gathered and ground the seeds that nourished our kin, danced the rhythms of the earth in the ceremonies by which the world is renewed, and sang the stories of our Ancestors as we moved through Country. Then came the strangers.

Aunty Joan Winch (Nyungar and Martujarra) once travelled back in time through a dream to witness the moment that colonizers arrived in Nyungar Country. She recalls: 'a terrible feeling of doom came into my heart and I said to myself, "This is the beginning of the end"' (2008: 228). And the colonial apocalypse was not one but many—a cataclysm of violence that began anew whenever colonizers arrived in the homeland of an Indigenous nation. In the southwest of Western Australia, in Nyungar Country, it began in the 1820s. In the north where the Country of the Palyku lies, it was the 1860s. The same broad patterns repeated across the continent and around the world. Judy Atkinson (Bundjalung and Jiman), drawing on the work of Donald Baker, has identified three types of violence that facilitate colonialism: overt physical violence (invasion, disease, death and destruction), covert structural violence (enforced dependency, legislation, reserves and removals) and psychosocial domination (cultural and spiritual genocide) (2002: 59–73).

These cycles included sustained sexual violence against Indigenous women and children inflicted over the course of generations (Atkinson 2002; Watson 2007). As this violence was inscribed upon the bodies of Indigenous people so too was it inscribed upon our Countries. As Irene Watson (Tanganekald, Meintangk Boandik) writes, 'The colonial frontier men raped both land and bodies and the violence of the frontier is the most significant colonial legacy Aboriginal peoples carry' (2007: 102). And the white women of the frontier were not innocent of this violence. In Atkinson's words: '[white] society viewed sexual violations of Indigenous women as familiar male sporting events. White women maintained their silence in their denial of the reality of this violence' (2002: 62). As the chaotic violence of the frontier shifted into the organized violence of the (so-called) 'protection' era, white women were very much a part of creating the Stolen Generations of Indigenous children. It was white women who were charged with managing domestic matters in the households into which Stolen girls were placed, and white women who not only worked in, but managed, some of the institutions where the children were incarcerated.

The *Bringing Them Home* report estimates that one in ten girls was sexually abused in an institution or work placement, and not all of this abuse was perpetrated by men (1997: np). This is not to deny the voices of those white women who spoke out at the time (e.g. Holland 2015) but nor is it to obscure the realities of colonialism and race privilege. As historian Anne Curthoys has noted:

> Feminist investigations of national identity in Australia and similar societies will find the history of dispossession, exploitation, racism and segregation to be fundamental, not peripheral, to their project. In doing so, they will need to revise an assumption which lies at the heart of much feminist scholarship—the historical innocence of women. (1993: 174)

In linear terms it has been over 200 years since colonizers came to Australia. Yet we are not so very far from colonization given its continuing effects on the lives of Indigenous peoples (see United Nations Permanent Forum on Indigenous Issues 2009). These effects include the intersecting complexities of trauma and oppression experienced by Indigenous women. As Pat Dudgeon (Bardi) writes:

> Aboriginal women's perceptions of identity, gender and sexuality are inextricably intertwined—and deeply rooted in the Australian colonizing project. As women, they have suffered racism unlike white women; as Aboriginal women, they have suffered sexism unlike Aboriginal men; and they have suffered misrepresentation of their traditional roles and misbeliefs that as women they are better off under the structures of Western colonization. (2017: 108)

Further, a linear distance of hundreds of years has little meaning in Indigenous systems—and Indigenous Countries—where time is cyclical. Time does not run in a line from the past through the present and on into the future but is simply an aspect of the movement of all life through cycles of creation. Time, like everything else, exists in space and is as susceptible to action and reaction as any other life. The measure by which anyone has moved 'beyond' an event is not therefore the passage of linear years but the extent to which affected relationships have been healed. To put this another way, the inhabitants of a colonized land are only ever as far from colonization as to the degree to which they have achieved decolonization, remembering always that decolonization 'is not accountable to settlers, or settler futurity ... [but to] Indigenous sovereignty and futurity' (Tuck and Yang 2012: 35).

There is the possibility of decolonizing dialogues between ecofeminists and Indigenous women. But there is also a danger that ecofeminism will reproduce both the failures of feminism to deal with intersectional oppression of Indigenous women or the complicity of settler women in this oppression (Moreton-Robinson 2000), and the failure of the environmental movement to recognize the value of Indigenous management of our cultured Countries. Further, it is my belief that ecofeminists who do not interrogate position and complicity will promulgate rather than challenge the structures that continue to oppress Indigenous homelands and Indigenous women. In this context, Aileen Moreton-Robinson (Goenpul) has characterized feminist standpoints (including ecofeminist standpoints) as incommensurate with those of Indigenous women for two primary reasons: first, because feminists subscribe to an earth/body split that does not exist in Indigenous worldviews, and second, because feminists 'do not address their privileged relationship to the nation's sovereignty that underpins their situatedness and ontology' (2013: 6).

In this sense, this chapter is concerned with a fundamental question that underlies all strands of ecofeminist thought: how can ecofeminists seek justice for women and the earth in a way that does not wreak *in*justice upon Indigenous women and Indigenous homelands? As Lindsay Nixon (Anishinaabekwe, Nehiyaw-iskwew) has written:

> If eco-feminists truly want to engage with Indigenous feminism to legitimize their own movements, they must first engage with their own positionality and privilege as settlers: a positionality on which the continuation of settler-colonialism and the ongoing genocide of Indigenous peoples are prefaced. (2015: n.p)

I suggest then that ecofeminists—and all non-Indigenous scholars with an interest in forging a just future—change the way in which they relate to Indigenous peoples and Indigenous homelands by respecting Indigenous sovereignty.

The Nature of Indigenous Sovereignty

Core to Indigenous notions of sovereignty is that any right to territory is intertwined with a responsibility to care for it. In relation to Australian Aboriginal nations I have previously described this as 'narrative sovereignty' whereby belonging is grounded in story and particularly in the stories of the Ancestor beings (2017: 9).

The homelands of Australian Aboriginal nations were formed in what is sometimes called the Dreaming, although this is an inexact translation of an Indigenous concept that has no English language equivalent. The Dreaming is the ongoing creation of all that is. The many Dreaming

Ancestors formed reality as we know it and continue to exist today in the world that they made. Through their stories, Aboriginal peoples connect to Country as family. Our grandfathers and grandmothers, aunties and uncles and brothers and sisters are in the trees, the rock, the birds, the wind, the sun and the moon. Our Countries are places where everything lives and everything connects in an ever-moving web of relationships, and it is through sustaining these connections that everything continues. This is why Aboriginal law, at its broadest, is the means of maintaining the connections that are the world. It follows that actions which damage or destroy relationships are generally unlawful, and Aboriginal systems recognize familial relationships with all forms of life.

Women existed within these networks of relationships and followed women's law. Among the responsibilities of women were caring for the women's spaces in Country, spaces to which we are related through the bloodlines that connect us to the Dreaming Ancestors. The femininity of Aboriginal women remains grounded in our culture and Countries. As Dawn Besserab (Bardi and Indjarbandi), writing of a study of Aboriginal women, has noted:

> Central to learning how to be women was learning about their Aboriginality. Women talked about learning to give and have respect for their elders, and about relationships and the importance of extended family in supporting and teaching them about their femininity. They learned from their grandparents, aunties and uncles concerning country, spirituality and the different responsibilities of caring for each other. (2017: 183)

The web of relationships that is Country repeats in every aspect—the pattern exists in a single grain of sand, and is formed again by millions of grains coming together to make desert; it is in spinifex and crow and rock and human and every other shape of life; and is created anew when these shapes come together to form Country. This means that harm to Country results in harm to Indigenous people; the reverse is also true. The wellbeing of Indigenous women and the wellbeing of our Countries—and especially of the women's places therein—are one and the same.

Sovereignty, like Country, is who we are.

Respecting Indigenous Sovereignty

It flows from the above that respecting Indigenous sovereignty requires far more than a simple acknowledgement that we were here before others came to our homelands. It requires a fundamental shift in the way non-Indigenous peoples orient themselves in the world. As white academic Fiona Nicoll has written, the critical question to ask is: 'What is the relationship of other Australians, in the name of whom national sovereignty is claimed and defended, to … Indigenous sovereignty?' (2004: np). This is a question that must be asked not once but many times over, a constant interrogation of an individual's place in a colonized land and their privilege in relation to colonized peoples. Respecting Indigenous sovereignty is a way of being and one which requires a focus on process.

Among the key differences between Indigenous and Western systems is that Indigenous systems tend to orientate towards process. This is a logical consequence of a holistic view of reality as comprised of many forms of life shifting through cycles of creation. On such a view, the sun will not rise and set, the rivers will not flow, the trees will not stretch their branches to the sky, and the birds will not sing their welcome to the morning just because these things happened yesterday or even for the hundred days before that. All of this happens because human beings play their part in keeping the world turning by engaging in the many processes that sustain the relationships that are Country. Thus, in a larger sense, the difference between a planet that sustains our species and one that destroys us (or that we destroy) is the processes by which we live our lives and influence the lives of others.

So the question becomes, what are the processes through which a respect for Indigenous women's sovereignty can be meaningfully realized by all those now living in Indigenous homelands? This chapter focuses on a single process—that of listening—but it is far from the only one. Thus, the discussion below is intended as a point of entry into a much larger journey.

Enacting Respect Through Listening

Nixon has written that 'What Indigenous feminists want from ecofeminists is simple: Sit down, be quiet, and listen' (2015: np). To listen is an act of transformative power. It is also one that is often misunderstood. Listening, in this context, does not mean the cultures and identities of Indigenous women are available for appropriation by ecofeminists, and this includes describing our lives as some idealized form of ecofemininity. On the contrary, the very nature of respecting sovereignty requires recognizing that our lives are for us to share (or not) on our own terms.

To truly listen to another requires that the person listening is not speaking. No one is able to listen to Indigenous voices whilst simultaneously engaged in the process of telling Indigenous women what we want, what is best for us and who we are. This practice of talking without hearing has frequently been a feature of the work of past researchers and policy-makers. It has also been a feature of feminism. In this context, Larissa Behrendt (Eualeyai and Kamillaroi) has commented that some feminist writers:

> are telling Aboriginal women not to see what they see: that their position in society is defined by their gender rather than their race, that the push for rights by white women will empower black women, that we are aligned with white women in the battle against oppression and that white women are as oppressed as we are. We do not believe any of these white lies. The experiences of black women are trivialized when viewed as merely an extension of the experiences of white women. (1993: 41)

So, to all ecofeminists with an interest in the subject of this chapter, I say: take a breath. Let go of what you know—or think you know—about Indigenous women, cultures and homelands. And listen instead to what we say about who we are.

Secondly, listening to Indigenous voices requires *Indigenous* voices. Too often, the words of others speaking about us and our realities—including those of non-Indigenous academics—have been listened to in place of our own. This has prompted a number of researchers, including myself, to privilege Indigenous voices in our work as the primary and most authentic sources of our own realities (Kwaymullina et al. 2013; Martin 2003; Rigney 1999). It is important as well to ensure that any information being drawn upon is knowledge that Indigenous women have chosen to share, rather than information that has found its way into the public domain through unethical research practices.

The use of ethically published sources will ensure ecofeminists are not violating the boundaries of Indigenous women through listening to knowledge that has been placed in the public domain without the permission of, or any return of benefits to, the Indigenous women knowledge holders. An appreciation of the considerations that govern ethical publishing can be acquired by reading the *Guidelines for the ethical publishing of Aboriginal and Torres Strait Islander authors and research from those communities* developed by Aboriginal Studies Press (Australian Institute of Aboriginal and Torres Strait Islander Studies 2015). There is now a vast array of ethically published Indigenous works available, with the majority of Australian Indigenous works produced by Indigenous small presses such as Magabala Books, IAD Press and Aboriginal Studies Press. Women's stories published by these presses include stories that speak to Indigenous law- and life-ways in Country (Turner 2010; Wallace 2009), stories of the ongoing battle for land justice (Kartinyeri and Anderson 2008; Lennon 2011), and stories of the struggles and triumphs of Indigenous women in the academy (Dudgeon et al. 2017).

Third, listening requires the capacity to hear the word 'no'. Historically, Western research has been a serial violator of the borders of Indigenous peoples. These violations have ranged from the physical (such as the taking of skin, hair and bones) to the ideological (the appropriation of Indigenous cultures and knowledges) (Smith 2012). And the presumed right of the scholars of the West to dispossess Indigenous peoples of our bodies, cultures and knowledges was grounded in, and sustained by, the same justification used to seize Indigenous territory: the characterization of Indigenous life-ways as inferior to those of Western Europe (Miller et al. 2010).

To respect Indigenous sovereignty therefore requires respecting the right of Indigenous women to share as much or as little of ourselves as we choose to. This is a principle enshrined in the *Declaration on the Rights of*

Indigenous Peoples (United Nations General Assembly 2007) and in contemporary ethical research protocols (Australian Institute of Aboriginal and Torres Strait Islander Studies 2012, principle 6) as 'free, prior and informed consent'. In a research context, 'free' means free from all coercion, manipulation and pressure, including the pressure of unrealistic timeframes within which to respond. 'Prior' means prior to a research project taking place or any change in the research. 'Informed' means being fully informed as to a range of matters, including the management of Indigenous Cultural and Intellectual Property and the return of benefits to Indigenous knowledge holders (United Nations Permanent Forum on Indigenous Issues 2005; Kimberley Land Council 2011). Non-Indigenous women scholars who fail to respect the need to ask first (and accept that the answer may be no) will themselves become trespassers across the boundaries of Indigenous women, promulgating some of the very injustices ecofeminists seek to challenge.

Fourth, all individuals must interrogate the preconceptions and the privilege that will prevent them from hearing Indigenous women. This means listening to Indigenous women on our own terms, rather than trying to fit our words and realities into Western knowledge frameworks or into what others think they 'know' of us. We are many voices speaking to many realities. Be challenged by us, puzzled by us, inspired by us—but allow us to be who we are unhindered by outside expectations of who we should be. And learn, too, the art of continually checking and challenging your own privilege. Settler women hold less power than settler men but vastly more than Indigenous women, and the marginalization of Indigenous women is the direct result of the dispossession from which all of settler Australia continues to benefit. In the words of Behrendt: 'Aboriginal women, who enjoyed power and respect within their traditional communities, fell to the lowest rung on the socio-economic ladder in colonial society because of the double taint of a subordinated race and a subordinated gender' (2000: 364). Part of interrogating privilege includes interrogating the knowledge disciplines within which scholars work—including all forms of feminism—and the whiteness (or Eurocentrism) of the discipline itself (Henderson 2000; Moreton-Robinson 2004, 2011; Scheurich and Young 1997; Watson 2014).

Finally, what happens at the end of listening? My answer to this is that there is no end to listening; it is part of a lifelong process of critical reflexivity. But for those who find value in Indigenous words, the next matter to consider is how to respond in a way that remains respectful of Indigenous sovereignty. This is a conversation outside the scope of this chapter. But in relation to scholarship, I note that one of the measures of respect is not necessarily in the words a non-Indigenous scholar has written about Indigenous peoples and our homelands but in how they have supported the right of Indigenous women to speak our own words or to maintain our silences. It is in the degree to which they have interrogated positionality, whiteness and colonialism, rather than interrogating Indigeneity. It is in how often they have challenged the absence or under-representation of Indigenous women's voices and, to this extent, is in the spaces they have contested—or yielded to Indigenous women—rather than in those they have occupied. In this context, Bronwyn Fredericks (Murri) has challenged feminism to interrogate its own colonizing behaviours:

> I have not witnessed or participated in a forum where the very essence of what constituted that feminist site or held that group together was up for discussion; also excluded have been questions about how the event or the organizers could be more inclusive, how we could all explore what we mean by feminism, or how the forum could be more open to participation by Aboriginal women. In effect, nonindigenous Australian women are saying 'This is ours' or 'This is mine' because their actions and nonactions demonstrate the possessive logic of white sovereignty. (2010: 547–548)

Conclusion

My mother once wrote of the day when the ships of the First Fleet breached the horizon, imagining 'two vastly different peoples …[gazing] at each other in curiosity across a brilliantly blue watery expanse'. She mused:

> The innocent, idealistic part of me likes to believe there is hope in every moment; that the die is never cast until the moment has past. Perhaps even then, despite the unlawful claiming of our land, hope lingered … Far

> seeing individuals are born into every nation. This causes me to believe that the possibility of forging a just future is always present, if only we have the vision, will and courage to pursue it. (2008: 274)

When time is measured in non-linear terms, Indigenous and non-Indigenous peoples are in many respects still living on the frontier. But to conceive of time in a non-linear way also means that the passage of linear years has never moved any of us so far that we cannot take meaningful action to heal the wounds of colonialism. The opportunity still exists for relationships founded in respect for all those with the vision, will and courage to pursue decolonization. In this regard, there is the possibility of a dialogue between Indigenous peoples and ecofeminism provided ecofeminists can shift to a way of being that embodies respect for Indigenous women's sovereignty. And out of such dialogues, just futures are born.

Bibliography

Atkinson, Judy. 2002. *Trauma Trails, Recreating Song Lines: The Transgenerational Effects of Trauma in Indigenous Australia*. Melbourne, Australia: Spinifex Press.

Australian Institute of Aboriginal and Torres Strait Islander Studies. 2012. *Guidelines for Ethical Research in Indigenous Australian Studies*. http://aiatsis.gov.au/research/ethical-research/guidelines-ethical-research-Australian-indigenous-studies

———. (2015). *Guidelines for the Ethical Publishing of Aboriginal and Torres Strait Islander Authors and Research From Those Communities*. http://aiatsis.gov.au/aboriginal-studies-press/getting-published/ethical-publishing-guidelines

Behrendt, Larissa. 2000. Consent in a (Neo)Colonial Society: Aboriginal Women as Sexual and Legal 'Other'. *Australian Feminist Studies* 15 (33): 353–367.

———. 1993. Aboriginal Women and the White Lies of the Feminist Movement: Implications for Aboriginal Women in Rights Discourse. *Australian Feminist Law Journal* 1: 27–44.

Besserab, Dawn. 2017. Interrogating Gender: What's Race and Class Got to Do With It? In *Us Women, Our Ways, Our World*, ed. P. Dudgeon, J. Herbert, J. Milroy, and D. Oxenham. Broome: Magabala Books.

Curthoys, Anne. 1993. Identity Crisis: Colonialism, Nation and Gender in Australian History. *Gender and History* 5 (2): 165–176.

Dudgeon, Pat. 2017. Mothers of Sin: Indigenous Women's Perceptions of their Identity and Gender. In *Us Women, Our Ways, Our World*, ed. P. Dudgeon, J. Herbert, J. Milroy, and D. Oxenham. Broome: Magabala Books.

Dudgeon, Pat, Jeannie Herbert, Jill Milroy, and Darlene Oxenham, eds. 2017. *Us Women, Our Ways, Our World*. Broome: Magabala Books.

Fredericks, Bronwyn. 2010. Reempowering Ourselves: Australian Aboriginal Women. *Signs* 35 (3): 546–550.

Henderson, James (Sakej). 2000. Challenges of Respecting Indigenous World Views in the Eurocentric Education. In *Voice of the Drum: Indigenous Education and Culture*, ed. R. Neil. Brandon, Canada: Kingfisher.

Holland, Alison. 2015. *Just Relations: The Story of Mary Bennett's Crusade for Aboriginal Rights*. Crawley, Perth: University of Western Australia Press.

Human Rights and Equal Opportunity Commission. 1997. *Bringing Them Home: Report of the National Inquiry into the Separation of Aboriginal and Torres Strait Islander Children from Their Families*. Sydney: Commonwealth of Australia.

Kartinyeri, Doreen, and Sue Anderson. 2008. *Doreen Kartinyeri: My Ngarrindjeri Calling*. Canberra: Aboriginal Studies Press.

Kimberley Land Council. 2011. *Kimberley Land Council Intellectual Property and Traditional Knowledge Policy*. http://www.klc.org.au/news-media/research-facilitation

Kwaymullina, Ambelin. 2017. Aboriginal Nations, the Australian Nation-State and Indigenous International Legal Traditions. In *Indigenous Peoples as Subjects of International Law*, ed. I. Watson. New York: Routledge.

Kwaymullina, Ambelin, Blaze Kwaymullina, and Lauren Butterly. 2013. Living Texts: A Perspective on Published Sources, Indigenous Research Methodologies and Indigenous Worldviews. *International Journal of Critical Indigenous Studies* 6 (1): 1–13.

Lennon, Jessie. 2011. *I'm the One That Know This Country!* Canberra: Aboriginal Studies Press.

Martin, Karen. 2003. Ways of Knowing, Being and Doing: A Theoretical Framework and Methods for Indigenous and Indigenist Research. *Journal of Australian Studies* 27 (76): 203–214.

Miller, Robert, Jacinta Ruru, Larissa Behrendt, and Tracey Lindberg. 2010. *Discovering Indigenous Lands: The Doctrine of Discovery in the English Colonies*. Oxford: Oxford University Press.

Moreton-Robinson, Aileen, ed. 2000. *Talkin' Up to the White Woman*. St. Lucia, Queensland: University of Queensland Press.

———. 2004. Whiteness, Epistemology and Indigenous Representation. In *Whitening Race: Essays in Social and Cultural Criticism*, ed. A. Moreton-Robinson. Canberra: Aboriginal Studies Press.

———. 2011. The White Man's Burden: Patriarchal White Epistemic Violence and Aboriginal Women's Knowledges Within the Academy. *Australian Feminist Studies* 26 (70): 413–431.

———. 2013. Towards an Australian Indigenous Women's Standpoint Theory. *Australian Feminist Studies* 28 (78): 331–347.

Morgan, Sally. 2008. The Balance for the World. In *Heartsick for Country*, ed. S. Morgan, T. Mia, and B. Kwaymullina. Fremantle: Fremantle Press.

Nicoll, Fiona. 2004. Are You Calling Me A Racist? Teaching Critical Whiteness Theory in Indigenous Sovereignty. *Borderlands* 3 (2). Accessed May 5, 2017. http://www.borderlands.net.au

Nixon, Lindsay. 2015. Eco-Feminist Appropriations of Indigenous Feminisms and Environmental Violence. *Feminist wire*. Accessed May 5, 2017. http://www.thefeministwire.com/2015/04/eco-feminist-appropriations-of-indigenous-feminisms-and-environmental-violence/

Oxenham, Darlene, Jeannie Herbert, Jill Milroy, and Pat Dudgeon, eds. 2017. *Us Women, Our Ways, Our World*. Broome: Magabala Books.

Rigney, Lester. 1999. Internationalization of an Indigenous Anticolonial Cultural Critique of Research Methodologies: A Guide to Indigenist Research Methodology and Its Principles. *Wicazo Sa Review* 14 (2): 109–121.

Scheurich, James, and Michelle Young. 1997. Coloring Epistemologies: Are Our Research Epistemologies Racially Biased? *Educational Researcher* 26 (4): 4–16.

Smith, Linda. 2012. *Decolonizing Methodologies: Research and Indigenous Peoples*. 2nd ed. London, England: Zed Books.

Tuck, Eve, and K. Wayne Yang. 2012. Decolonization is Not a Metaphor. *Decolonization: Indigeneity, Education & Society* 1 (1): 1–40.

Turner, Margaret Kemarre. 2010. *Iwenhe Tyerrtye: What It Means to Be an Aboriginal Person*. Alice Springs: IAD Press.

United Nations General Assembly. 2007. *United Nations Declaration on the Rights of Indigenous Peoples*. A/RES/61/295. October 2.

United Nations Permanent Forum on Indigenous Issues (UNPFII). 2010. *State of the World's Indigenous Peoples*, 2010, 14 January. ST/ESA/328. Accessed May 21, 2017. http://www.refworld.org/docid/4b6700ed2.html

United Nations Permanent Forum on Indigenous Issues. 2005. *Report of the International Workshop on Methodologies Regarding Free, Prior and Informed Consent and Indigenous Peoples*, January 17–19, E/C.19/2005/3.

Wallace, Kathleen Kemarre. 2009. *Listen Deeply: Let These Stories In*. Alice Springs: IAD Press.

Watson, Irene. 2007. Aboriginal Women's Laws and Lives: How Might We Keep Growing the Law? *Australian Feminist Law Journal* 26: 95–107.

———. 2014. Re-Centring First Nations Knowledge and Places in a Terra Nullius Space. *AlterNative* 10 (5): 508–520.

Watson, Lilla. 2004. The Recognition of Indigenous Terms of Reference. Keynote Address at *A Contribution to Change: Cooperation out of Conflict Conference: Celebrating Difference, Embracing Equality*, Hobart, September 21–24.

Whittaker, Alison. 2016. *Lemons in the Chicken Wire*. Broome: Magabala Books.

Winch, Joan. 2008. A Feeling of Belonging. In *Heartsick for Country*, ed. S. Morgan, T. Mia, and B. Kwaymullina. Fremantle: Fremantle Press.

Ambelin Kwaymullina is an Aboriginal law academic who comes from the Palyku people of the Pilbara region of Western Australia. Her research interests include Indigenous Cultural and Intellectual Property, Indigenous law- and life-ways, and ethical engagement with Indigenous peoples. Ambelin is also an award-winning children's writer and illustrator whose books have been translated into several languages.

12

Feminist Ecologies in Religious Interpretation: Australian Influences

Anne Elvey

As this volume attests, internationally significant Australian publication in ecological philosophy can be traced to at least as early as 1973 with Val and Richard Routley's (later Val Plumwood and Richard Sylvan) book *Fight for the Forests*. By the 1990s, Australian ecological and ecological feminist philosophy was at the forefront of international scholarship in this area, with Plumwood offering a careful and incisive analysis of dualism (or hyper-separation) as a logic of colonization (1993: 41–68). Although the term *écofeminisme* was likely coined by French feminist Françoise d'Eaubonne in 1974 to express the interconnectedness between oppression of women and ecological destruction, I suggest that ecological feminism did not appear explicitly in biblical interpretation until the 1990s, and remains marginal in the field of biblical studies (see also Elvey forthcoming a, b).

This chapter charts a recent history of feminist interpretation of biblical religion and an evolving, but uneven, relationship between feminist

A. Elvey (✉)
Monash University, Clayton, VIC, Australia
University of Divinity, Melbourne, VIC, Australia

L. Stevens et al. (eds.), *Feminist Ecologies*,
https://doi.org/10.1007/978-3-319-64385-4_12

and ecological thinking in biblical studies. While ecological thinking in biblical religion was expressly manifest in Australia in the 1990s (Byrne 1990; Kelly 1993; Edwards 1995), a feminist influence in ecological biblical interpretation was less evident until the emergence of the Earth Bible Project in Adelaide, South Australia, with its first major publications appearing in 2000. This Project was also influenced by indigenous cultures particularly in Australia. The active work and publications of Elaine Wainwright, a contributor to the Earth Bible Project, also demonstrate the influence of indigenous perspectives from Australia, Aotearoa (also called New Zealand), and Oceania, on the development of ecological feminist hermeneutics in biblical religion. The first half of this chapter describes the early history, and the second part explores Wainwright's writings as indicative of how the postcolonial contexts in Australian and New Zealand have evolved some complex intersections of religion, race, and ecological thought.

Feminism and the Bible

First-wave feminism in North America gave rise to *The Woman's Bible*, a feminist rereading of biblical texts, under the editorship of Elizabeth Cady Stanton (1993 [1895–1898]), who saw biblical criticism as essential for women's liberation and argued that the Bible reflected the words of *men* as much or more than it did any divine word. While recognizing that Christian women within the Abolitionist struggle became, through the movement, aware of their own oppression under patriarchy, Stanton was critical of the exclusion of women from 'universal' suffrage. She acknowledged support for women's suffrage from Abolitionist and former slave Frederick Douglass (Stanton 1897). Such tensions and connections between women's suffrage and antislavery movements resonate in twentieth-century engagements between liberal feminist and other liberationist and justice movements. Similar tensions appear in critical conversations between Jewish and Christian feminist biblical scholars and in postcolonial biblical studies. For feminist biblical scholars, complexity surrounds not only the exclusion of women's voices and experiences beyond a dominant white, liberal majority in the Northern

Hemisphere, but also the ambiguity of reception and use of that collection (or book) of books, the Bible, that has sometimes inspired emancipation.

While many feminists rejected biblical religion as irredeemably patriarchal (Daly 1985; Hampson and Ruether 1987), second-wave feminism in North America also gave rise to now classic texts such as Elisabeth Schüssler Fiorenza's *In Memory of Her: A Feminist Theological Reconstruction of Christian Origins* (1983), Rosemary Radford Ruether's *Sexism and God-Talk: Toward a Feminist Theology* (1983), and Phyllis Trible's *God and the Rhetoric of Sexuality* (1978). These publications were concerned with rethinking the maleness of the divine, the construction of women, and the recovery and reconstruction of elided female histories in biblical religion. The wider context involved feminist reclamation of female deities (e.g., Starhawk 1989 [1979]; Gimbutas 1974) and the concept of a divine becoming in the feminine (Irigaray 1986). The reconstructive reclamation of the Goddess was a recovery of a woman-nature affiliation counter to the patriarchal denigration of both, but potentially reinscribing the problems of identifying women with nature, and continuing to leave men outside nature (see Rigby 1998: 151, 2001). In biblical religion, feminist reclamation of female divinities raised interest in goddesses such as Asherah and the figure of divine Wisdom (in Hebrew, Hokmah; in Greek, Sophia) (e.g., Brenner 1992; Hadley 1997).

Trible writes, '[b]y feminist I do not mean a narrow focus upon women, but rather a critique of culture in the light of misogyny' (1978: 7). Trible (1984) also critiques what she calls 'texts of terror'—biblical narratives of violence toward women—addressing the stories of Hagar (Gen 16:1–16; 21:9–21); Tamar (2 Sam 13:1–22); an unnamed woman raped, tortured, murdered, and dismembered (Judg 19:1–30); and the daughter of Jephthah, sacrificed because of her father's vow (Judg 11:29–40). Later, Norman Habel (2009, 2012) extends this notion to ecological texts of terror (e.g., Gen 6:11–13) and to a memorialization of contemporary scorched places. These critical interpretations become modes of memorial that both wound the interpreter and open to a shift in culture. Both cultural critique and an orientation to cultural change are crucial in ecological feminist work (Plumwood 2002).

Around the time Plumwood's *Feminism and the Mastery of Nature* (1993) appeared, Schüssler Fiorenza published her critical analysis of kyriarchy, a term she coined to express 'a different understanding of patriarchy, one which does not limit it to the sex/gender system but conceptualizes it in terms of interlocking structures of domination [i.e., kyriarchal, elite male, relations of ruling (*Herr-schaft*)]' (1992: 7–8). The term kyriarchy had an explicit link to biblical literature through the word *kyrios* (lord), both paterfamilias and master of slaves, also used as a title for the divine. Also published in 1992, Radford Ruether's *Gaia and God: An Ecofeminist Theology of Earth Healing* linked human violence, religious symbols, and ecological destruction, so that patriarchy and biblical religion were reexamined not only for their domination of women but equally for their Earth-destroying impacts, with a focus on healing Earth.

Schüssler Fiorenza's (1983) influential early work in biblical studies argued that all representations of Christian origins are constructions, so that feminists need to reconstruct such origins in ways that are empowering for women. In contrast to Plumwood, Schüssler Fiorenza's (1992) focus on kyriarchy did not extend beyond human dominance of other humans. Feminist reconstructive projects listened for the voices, experiences, and agencies of women. Similarly, current ecological reconstructive projects harken to the voices, experience, and modes of being and behaving of other than humans. But the difference is immense, in that human languages themselves are unsettled by the otherness of the communications of other than humans. In an ecological frame, Deborah Bird Rose and her coauthors explain that it is important, 'to resituate the human within the environment, and to resituate nonhumans within cultural and ethical domains' (2012: 3).

The Earth Bible Project

Twentieth-century Christian writing in response to ecological destruction can be traced at least as early as American Lutheran theological Joseph Sittler's *The Care of the Earth* (2004 [1964]). The question of relationship between biblical religion and environmental crisis came to prominence with the publication of Lynn White's 1967 essay, 'The

Historical Roots of the Ecological Crisis', which identified the attitudes and worldviews of biblical religion, particularly in its traditions of interpretation, as a root cause of ecological destruction. As I discuss elsewhere, a majority of biblical scholars ignored the critique; for them contextual approaches to the biblical text which focused on patriarchy, ecological destruction, or colonial oppression were of secondary importance to the study of the meaning of the biblical text itself. Some scholars, however, adopted and expanded White's criticism even to the extent of seeing the Bible as effectively toxic to Earth. More often, however, the response was apologetic, and scholars were at pains to highlight eco-friendly parts of the biblical corpus (Elvey 2005: 15–16).

By the 1990s, several Australian biblical scholars and theologians were beginning to hear an Earth 'cry' so to speak, and were writing on occasion with an ecological focus (see Byrne 1990; Kelly 1993; Edwards 1995). As the decade neared a close, the first major international collaborative research project in ecological hermeneutics in biblical studies, the Earth Bible Project, took shape in Adelaide, South Australia, under the chief editorship of Norman Habel. The project team was in conversation with feminists, ecologists, and a number of indigenous peoples from Australia and overseas, and produced a succession of key publications in the field. First, they developed six ecojustice principles: intrinsic worth, interconnectedness, purpose, voice, mutual custodianship, and resistance, and then three further ecological hermeneutics—suspicion, identification, and retrieval (Habel 2000, 2001; Habel and Wurst 2000, 2001; Habel and Balabanski 2002; Habel and Trudinger 2008).

In the context of the Earth Bible Project, Heather Eaton warns that '[f]or many, to read the Bible from an anti-woman and anti-Earth perspective is to enjoy a comfortable read' (2000: 70). In the third Earth Bible volume, Laura Hobgood-Oster highlights the ambiguous promise of biblical Wisdom traditions for women or Earth; then in counterpoint she emphasizes the agency of Earth and its capacity to 'subvert' and 'recontextualize' readings of the biblical texts themselves (2001: 45–46). The Earth Bible principles of voice—'Earth is a subject capable of raising its voice in celebration and against injustice'—and resistance—'Earth and its components not only suffer from injustices at the hands of humans, but actively resist them in the struggle for justice'—affirm

Earth's agency (Habel 2000: 24). Scholars applied these principles to readings that recovered or reconstructed biblical texts from Earth perspectives (Habel 2000; Habel and Trudinger 2008).

Vicky Balabanski, an Adelaide-based biblical scholar and member of the Earth Bible Team, sees the Earth Bible Project as 'a trajectory of feminist scholarship' (2007: 146). She writes: 'My own trajectory of feminist/post-patriarchal scholarship has found expression by means of the "Earth Bible" project' (146). The project had a collaborative strategy, she recalls, and aimed at reading 'the Bible in solidarity with the Earth, which is defined as an inclusive term encompassing the whole web of life, the total ecosystem, of which humanity is a part' (146).

Over the first decade of the twenty-first century, the Earth Bible eco-justice principles were refined especially through the Ecological Hermeneutics Program Unit of the Society of Biblical Literature (SBL) Annual Meeting, a major conference of biblical scholars from the United States and across the world, held over four days in November each year in the US. SBL also holds an annual International Meeting in July-August outside the US, and for several years the international meeting also had an Ecological Hermeneutics Program Unit, but no longer does. This chapter turns now to one of the key members of the Ecological Hermeneutics Program Unit, to explore the way her feminist and ecological interpretive practices have developed over 25 years.

Wainwright: Feminism and Ecology to Multidimensionality

In 1991, Elaine Wainwright, an Australian biblical scholar, feminist theologian, and Sister of Mercy, published an article, 'A Metaphorical Walk through Scripture in an Ecological Age' (1991b), although for many years her feminist work was more prominent than her ecological work. She admits that feminism is 'a relative newcomer' to the scholarly world especially in biblical studies (Wainwright 1991a: 9). Over the next two decades, Wainwright's published work assisted not only with the consolidation of ecological interpretations within biblical studies, but she was able to bring these into dialogue with postcolonial implications.

Crucial for any ecological reading praxis in colonized spaces such as Australia or New Zealand has been (and continues to be) a postcolonial—or, better, decolonizing—ethic. This possibility subsequently emerged for Wainwright in the context of her professional work and appointment as inaugural head of school and professor of theology at the University of Auckland. She held this role from 2002 to 2014 where she worked with Maori and Pacific Islander colleagues. Their exchanges were formalized in the Oceanic Biblical Studies Association which first gathered in 2010 in Auckland. Wainwright has been active in solidarity with women theologians of the Pacific and, in addition, as president of the Society for Asian Biblical Studies. This cross-cultural collegial praxis across Asia and the Pacific has shaped the multidimensional nature of her hermeneutics (Wainwright 2005).

Wainwright became part of the global academy of feminist and biblical scholars, in which she is both a mentor and conversation partner to many, and actively responsive to postcolonial and other critiques of her work. For example, African biblical scholar Musa Dube comments on Wainwright's first book that her 'gender inclusive reading is a remarkable feminist achievement in its clarity, its creativeness, and its persuasiveness, for it foregrounds the centrality of women to Matthew's story without underplaying the patriarchal and androcentric cores that contain them' (2000: 177). In Australia, Anne Pattel-Gray (1995, 1999) has challenged non-indigenous feminist theologians to recognize their own colonial 'social location' and privilege (Wainwright 1998: 9).

The origins of Wainwright's dialogue with representatives from disenfranchised communities who are particularly affected by environmental problems are evident in her publications from 'A Metaphorical Walk' in 1991. It opens with a poem by Australian Aboriginal poet, Oodgeroo, called 'Time is Running Out' which refers to the violence of mining as rape of the Earth (Wainwright 1991b: 273). Wainwright then reflects on the 'rate of extinction of species' and speaks of an 'urgent call coming from the earth itself' (274). The article develops an ecological context in relation to the cosmological focus of Thomas Berry, whose exposition of a New Story in contrast to the Old Stories of biblical religion and modernist mechanistic science remains popular today, especially among Roman Catholic religious, female and male.[1] Wainwright's article goes on to suggest a mode of ecological reading of texts modeled on feminist and liberationist approaches, appealing in particular to Schüssler Fiorenza's feminist hermeneutics of suspicion and reclamation. In a feminist frame the former applies to biblical texts a suspicion of androcentric perspectives and language and patriarchal interests; the latter, using creative imagination, shapes a 'new' story in dialogue with metaphors and images recovered from the biblical text. Wainwright's 1991 'Walk' is wide-ranging, critiquing progress and authority, separateness and territoriality and championing interconnectedness and interrelatedness. Drawing on Carolyn Merchant (1990 [1980]), Wainwright comments that 'at the heart of ecofeminism lies the principle of *inclusion*' (1991b: 281, emphasis in original; see also, 1991a: esp. 30–32).[2]

The second part of Wainwright's 'Walk' applies hermeneutics of suspicion and reclamation across a range of biblical texts and symbols, picking up on the anthropocentrism of Genesis 1; the possibilities in the practices of Sabbath for resting the land; the centrality of the land in the Hebrew Bible; the creative spirit of the female figure of Wisdom; the table companionship of Jesus; and the central symbols of cross and resurrection as critiquing 'any understanding of God or humanity in hierarchical or dominant terms' (1991b: 292). She then suggests the work of a diverse range of thinkers such as: Hildegard of Bingen; Daly River theologian, teacher, and artist Miriam Rose Ungunmerr; poet Judith Wright; and curator Rosemary Crumlin, as other sources, in conjunction with contemporary ecological and cosmological science, for subverting anti-ecological thinking and practice. Wainwright's article concludes: 'we must engage in revolutionary praxis to stem the tide of time which is running out and shape a future yet to be imagined' (294).

While Wainwright's work in the 1990s had an occasional reference to ecological reading (e.g., 1997), it retained a predominantly feminist focus, with a sense of the possibilities for transformative reading inhering in the plurality of any text. Wainwright considered patriarchy as 'a multidimensional system' including 'humanity's domination of [with some hyperbole] the universe' (1998: 2). Describing a poetics, rhetorics, and politics of engendered reading, Wainwright (1998) occasionally brought ecological concerns into conversation with readings of the text that focused on genre, gender, and engendering new meaning. Noting that the opening to the Gospel of Matthew recalls the Book of Genesis, and its genealogy of Earth and cosmos, Wainwright suggests that the rupturing of the patrilineage may also be an opening to an ecological kinship in response to ecological crisis (1998: 58).

Within the Earth Bible Project publications, Wainwright (2000, 2002) contributed essays to the first and fifth volumes, where she applied her deconstructive and reconstructive hermeneutics learned from feminism, together with a focus on voices of Earth and cosmos as voices resistant to the anthropocentrism of most biblical texts. She continued to read ecofeminist possibilities in the intertextual echoes of the feminine biblical figure of divine Wisdom. In the opening decades of the twenty-first century, Wainwright has been a leading contributor to the Ecological Hermeneutics Section of the Society of Biblical Literature Annual and International Meetings where the Earth Bible ecojustice principles were nuanced by the development of ecological hermeneutics of suspicion, identification, and retrieval, echoing but extending earlier feminist hermeneutics.

A Material Turn

In my ecological feminist reading of the Gospel of Luke, I draw on Gayatri Spivak's description of the pregnant body as 'prepropriative', and in Jacques Derrida's (1994) terms 'aneconomic', to explore critically in the biblical text a logic of gestation (Elvey 2005; Spivak 1993: 148; Derrida 1994: 7). This interpretation does not equate women with nature or Earth, rather it points to where the pregnant body, like the Earth, is necessary for many forms of animal life (Elvey 2003). At this point of similarity, pregnant bodies and Earth share a logic, and a mode, of being (a material givenness) that, already relational, unsettles distinctions between self and other and potentially resist problematic economies of exchange, especially such as occur within systems of capitalist consumerism (Elvey 2003, 2005). My subsequent work dovetails with new materialisms and also focuses on the materiality of texts as an important aspect of the ecological embeddedness of texts, their interpreters and their interpretations (Elvey 2004, 2010, 2011a, b). Wainwright and I have been in dialogue in relation to this emerging focus on materiality which also informs her work (2006, 2012, 2016).

Wainwright (2006) incorporates an ecological hermeneutic when she traces the work of women as healers and subjects of healing in antiquity and applies this to her reading of women and healing in the Gospels of Matthew, Mark, and Luke. She develops a multidimensional hermeneutic in which feminist, postcolonial, and ecological aspects are intertwined. She builds on the work of philosophers such as Plumwood, feminist biblical scholars such as Schüssler Fiorenza, and her own earlier insights into the interrelatedness of patriarchal, kyriarchal, and hyperseparatist systems of oppression. For the ecological aspect of her multidimensional hermeneutic, Wainwright highlights attention to the material, 'to the actualities and the shifts and changes in the physical realities of women's lives in antiquity', including attention to the herbs, ointments, and oils used in healing practices; attentiveness to Earth as supplier of these material resources for healing; attentiveness to the body; and a consideration of space, particularly its colonization and gendering (2006: 18–23, see also 2003).

The focus on the materiality of healing is most pronounced in Wainwright's treatments of the various gospel versions of the woman anointing Jesus. Unlike most gospel stories apart from the narratives of Jesus' death, there are four versions of the story of a woman anointing Jesus, one in each of the canonical gospels (Matt 26:6–13; Mark 14:3–9; Luke 7:36–50; John 12:1–8). While the four stories carry significant differences of detail and emphasis, they share many commonalities, in particular the key elements of a woman, an anointing, ointment, and Jesus as the one who is anointed, as well as critical onlookers. To highlight the materiality and material agency of the ointment, Wainwright titles one paper, 'Unbound Hair and *Ointmented* Feet' (2008b; emphasis added). Moreover, in this article on Luke 7:36–50, a new thread enters the weave with an ecological focus on the woman's erotic actions as transgressive and both expressive of the principle of interconnectedness and potentially unsettling of the hyperseparations Plumwood (2002) describes (Wainwright 2008b: esp. 182–184, see also 2015).

An article on the anointing in Mark 14:3–9 focuses on the hermeneutic of identification—'not [to] collapse difference but [to] allow for it within relationship' through a 'strong sympathetic or imaginative bond' (Wainwright 2008a: 132). She proposes to identify with the *muron*, the ointment of healing, in its materiality, its being identified as a commodity, and the multiple interdependencies that lie behind its characterization in the text as 'waste'. Drawing on British feminist ecotheologian Anne Primavesi's notion of gift, Wainwright depicts the narrative of healing as a 'gift event', in excess of the economies that commodify and waste, rendering some poor and others rich. Primavesi's (2000, 2003) conception of gift lacks the deconstructive edge of Derrida's 'aneconomic' gift but offers a Gaia systems-oriented focus on gift as event. This might be extrapolated toward a material 'grace', or a gracious materiality (cf. Plumwood 2002; Rigby 2014). In Wainwright's (2008a) article, the material agency of the *muron* is not yet fully articulated.

Attention to materiality and material space, noted above, however, becomes subtler in Wainwright's later work where she takes up the notion of habitat (2009, 2012, 2013). 'Habitat and in-habitants (the more-than-human)', writes Wainwright, 'are inseparable such that "habitat" can function as a key interpretive lens for reading ecologically' (2012: 293). Habitat includes place but is more than place. Lorraine Code focuses on 'habitat as a place to know'; she includes 'the social-political, cultural, and psychological elements … alongside physical and (other) environmental contributors to the "nature" of a habitat and its inhabitants, at any historical moment' (Code 2006: 37; see also Wainwright 2012: 298). The linking of habitat and knowledge is particularly important. Humans and other creatures come to know what is sustaining, for example, in relation to the environment in which they live; this environ, or habitat, is not simply the geographic place but complex relationships of climate, sources of food, clothing and shelter, the power relations that affect these factors, and much more. Moreover, habitat both shapes and is shaped by humans (Wainwright 2012: 298). Habitat itself exercises material effect, indeed agency.

The exploration of 'habitat' in the reading of biblical texts is at least threefold: (1) the text encodes a habitat or habitats, often ignored by interpreters, which can be brought to the fore in ecological readings; (2)

the reader is always already embedded in a habitat or habitats; (3) a focus on habitat suggests an orientation toward 'co-*habitat*-ion' as ecological praxis (Wainwright 2012: 293). Wainwright (2013) employs this concept of habitat in the first of these three ways, in a reading of Matthew 15:21–28. In this text a Canaanite woman approaches Jesus asking for healing for her daughter, and a verbal contest ensues between the woman and Jesus with reference to dogs. Wainwright explores ways in which the agency and power of the hinterland of Tyre, the bread, and the dogs 'push up' (borrowing a thought from Rigby 2004) through the narrative, its setting, and its 'metaphoric referent[s]' (2013: 116). In a later work, while suspicious of the use of animals as metaphors, and drawing on Donna Haraway (2008), Wainwright (2016) evokes an intersection of respect and artistry, when other beings are used as symbols in a text. She describes the 'carnal intertwining of sheep and people' (2016: 119). Reading the story of the Canaanite woman, she writes: 'Here, both Jesus and the woman have voice, bread links Tyrian and Jew and comes to represent the power to heal and *dogs, in their leaning toward their own materiality*, can enable Jesus to negotiate his own internal struggle to determine what is "word of God" and what is "human tradition" (15.6–9)' (162, emphasis added).

Wainwright (2016) adopts a global or planetary understanding of Earth as habitat for humans, where Earth has its origins in the cosmos and humans live in (not on), are nurtured, mentored, and transformed by Earth in all its complex and diverse interrelatedness of presences, beings, and voices. Her aim is to read as an Earth being from within the habitat of Earth and to invite other biblical scholars and interested readers to undertake similar readings on her model. Local or given habitats, in their variety of aspects, influence readers' identities before and as they come to read a text. Moreover, the multiplicity of habitats that influence reading and the complexity of the human cultures and histories that shape the situations of readers imply that ecological and postcolonial understandings of relationship to/in place are entwined. Further, habitat has more-than-human connotations; in relation to the text, the task is not only to identify, but more particularly to identify with, the other-than-human creature and/in its habitat; it is this habitat which could be said to shape the 'voice' of the other creature much as human habitats shape human voices.

Such identification is at best an imaginative approximation to the experience of the other, given the potential for recolonizing the other. It is what Mark Brett describes (in another context) as 'kenotic hospitality' (2008: 197). For Rigby, engaging with the biblical flood narratives and Noah's ark—in a Derridean frame of 'unconditional hospitality' that extends to the other animal—such hospitality might be understood as counter-utopian (2008: 173; Derrida 2000: 25). In this frame, the ark (that uncanny and perhaps paradigmatic post-utopian habitat) 'models a form of "ecstatic dwelling", dwelling, that is, in exile in the company of more-than-human strangers, which is fast becoming the only kind of dwelling available on an increasingly uninhabitable earth, where ever more beings, human and otherwise, are destined to be rendered homeless' (Rigby 2008: 174).

The potential impact on the communities of the majority (sometimes called third) world makes this issue urgent. As Dube asks of Wainwright, have the imperialist aspects of the interpretation of the biblical text been recognized and critiqued sufficiently (2000: 179)? More generally, is a feminism focused on gender sufficient for a postcolonial feminist project of decolonization of imperial biblical ideologies? Following earlier critiques of Christian feminist anti-Judaism (e.g., Plaskow 1991), Jewish feminist New Testament scholar, Amy-Jill Levine (2004a, b), critiques aspects of Christian feminist and postcolonial biblical scholarship, for its unconscious reinscription of a Christian supersessionism, that is, the view that Christianity is superior to and has superseded or replaced Judaism, especially problematic in light of the violent history of anti-Judaism. In response to Jewish, indigenous, and postcolonial critiques, Wainwright has redeveloped her feminist hermeneutics beyond its liberal beginnings toward the multidimensional approach that interweaves feminist with postcolonial and ecological hermeneutics, and which privileges a form of ecological thinking (e.g., 1995, 1998: 122 n. 8, 2016: 37).

Conclusion

For Wainwright, ecological reading, like feminist reading, is oriented toward culturally transformative praxis. She writes:

> Situating ecological reading within the context of ecological citizenship as a way of being in the world, of being a participating Earth-being, means that the end of this theoretical consideration is not an end nor a beginning but participates in the ongoing praxis of living ecologically responsible lives. Images, words, and stories can be transformative; they can arise from the Earth's body, but they will do so within the web of emerging ecological thinking as a new way of in-habiting Earth. (Wainwright 2012: 304)

Given Christian discourse of the 'new', especially in the notion of 'New' Testament and the eschatological hope of a 'new' or renewed creation, it is unsurprising that as a Christian feminist biblical scholar, Wainwright also refers to the 'new': a new story, a new way, for responding to the current moment. The appeal to the 'new' is commonplace in ecological spirituality broadly understood. Drawing on the biblical notion of *kairos*—a critical season, also the proper time—and on ecotheological applications of this notion to our own time, Rigby is wary of the 'idealist illusion that all we need is a "new story" to budge the entrenched socioeconomic and power-political interests that are keeping us on the path to catastrophe' (2015: 177). As Rigby writes, narratives alone are insufficient for forming attitudes and catalyzing actions; rather 'narratives and practices coconstitute one another' (2015: 177). For Wainwright the work of reinterpreting ancient texts is part of a commitment to contemporary more-than-human communities and their habitats.

Wainwright critiques theologies that focus on the 'truths of the past' and argues for context-based work, by communities of interpretation, where practices of interpretation shape communities (2005: 125–26). She writes that the 'complexity of our "being in the world" is recognized when one begins to examine the varieties of interpretive worlds constructed through memories, feelings, imaginations, thinking and action …' (125). In her work, especially through her scholarly engagement with colleagues, students and movements of women in and beyond the churches, especially in Australia, Aotearoa (New Zealand), and Oceania, Wainwright has developed her hermeneutic practice from a predominantly feminist ethic with occasional forays into ecological thinking to a genuine multidimensional approach (with entwined feminist, ecological, materialist, and postcolonial ethics), recognizing that ecological thinking is not simply an extrapolation of feminist or liberationist thinking but requires a shift of perspective (2016: 73).[3] The current volume's focus on feminist ecologies rather than ecological feminism makes a strong claim for the feminist roots of, and in, ecological thinking and practice. As I have shown in relation to Wainwright's work, this link is evident in ecological biblical studies—both her own and the Earth Bible Project. But there are tensions, and ecological thinking and practice require a more-than-human frame of reference that challenges feminists to resist our own anthropocentrism.

Notes

1. In Melbourne such interest is represented by groups such as Earthsong (http://earthsong.org.au/) and the Ecology and Spirituality Centre, Glenburn (http://www.edmundrice.org/glenburn.html). In some respects, there is a gap between popular movements in ecological spirituality and academic work in ecological theology and biblical studies, similar to the gap between grassroots feminist theology and feminist theologians in Australia.
2. This principle, developed by Wainwright in a feminist context, is later taken up by Mercy sisters and biblical scholars Elizabeth Dowling and Veronica Lawson (2013) to suggest that women's inclusion as Eucharistic leaders is implied in the inclusive principle of creation presented at the end of the Gospel of Mark.
3. In 2015 Wainwright took up a position as Executive Director of Mission and Ministry, Institute of Sisters of Mercy Australia and Papua New Guinea, a position with a strong social and ecological justice focus. My chapter deals only with a selection of Wainwright's work. For a recent curriculum vitae and selected publications list, see http://www.mercyworld.org/_uploads/projects/119-a30eaef1/user-assets/files/CV/CV_MIRP_E-Wainwright(1).pdf

Bibliography

Balabanski, Vicky. 2007. Ecological Hermeneutics as a Daughter of Feminism: Reflections on the Earth Bible Project. *Women-Church: An Australian Journal of Feminist Studies in Religion* 40: 145–149.

Brenner, Athalya. 1992. The Hebrew God and His Female Complements. In *The Feminist Companion to Mythology*, ed. C. Larrington, 48–62. London: Pandora Press.

Brett, Mark G. 2008. *Decolonizing God: The Bible in the Tides of Empire*. Sheffield: Sheffield Phoenix Press.

Byrne, Brendan. 1990. *Inheriting the Earth: The Pauline Basis of a Spirituality for Our Time*. Homebush, NSW: St. Paul Publications.

Code, Loraine L. 2006. *Ecological Thinking: The Politics of Epistemic Location*. New York: Oxford University Press.

Daly, Mary. (1973) 1985. *Beyond God the Father: A Philosophy of Women's Liberation*. With an Original Reintroduction by the Author. Boston: Beacon Press.

Derrida, Jacques. 1994. *Given Time: I. Counterfeit Money*. Trans. P. Kamuf. Chicago: University of Chicago Press.

———. 2000. *Of Hospitality. Anne Dufourmantelle Invites Jacques Derrida to Respond*. Trans. R. Bowlby. Stanford: Stanford University Press.

Dowling, Elizabeth, and Veronica Lawson. 2013. Women, Eucharist, and Good News to All Creation in Mark. In *Reinterpreting the Eucharist: Explorations in Feminist Theology and Ethics*, ed. A. Elvey, C. Hogan, K. Power, and C. Renkin, 78–90. Sheffield: Equinox Press.

Dube, Musa W. 2000. *Postcolonial Feminist Interpretation of the Bible*. Danvers, MA: Chalice Press.

Eaton, Heather. 2000. Ecofeminist Contributions to an Ecojustice Hermeneutics. In *Readings from the Perspective of Earth*, ed. N. C. Habel, 54–71. Earth Bible, 1. Sheffield: Sheffield Academic Press.

Edwards, Denis. 1995. *Jesus the Wisdom of God: An Ecological Theology*. Maryknoll, NY: Orbis Books.

Elvey, Anne F. 2003. The Material Given: Bodies, Pregnant Bodies and Earth. *Australian Feminist Studies* 18 (4, July): 199–209.

———. 2004. Earthing the Text? On the Status of the Biblical Text in Ecological Perspective. *Australian Biblical Review* 52: 64–79.

———. 2005. *An Ecological Feminist Reading of the Gospel of Luke: A Gestational Paradigm*. Studies in Women and Religion 45. Lewiston: Edwin Mellen Press.

———. 2010. The Bible and the Earth: Toward a Materialist Ecohermeneutics. *Colloquium: The Australian and New Zealand Theological Review* 42 (2, Nov.): 131–150.

———. 2011a. *The Matter of the Text: Material Engagements Between Luke and the Five Senses*. The Bible in the Modern World 37. Sheffield: Sheffield Phoenix Press.

———. 2011b. The Matter of Texts: A Material Intertextuality and Ecocritical Engagements with the Bible. In *Ecocritical Theory: New European Approaches*, ed. A. Goodbody and K. Rigby, 181–193. Charlottesville: University of Virginia Press.

———. forthcoming a. Ecological Feminist Hermeneutics. In *The Oxford Handbook of the Bible and Ecology*, ed. H. Marlow and M. Harris. Oxford: Oxford University Press.

———. forthcoming b. Some Possibilities and Tensions of Feminism and Ecological Criticism. In *The Oxford Handbook of Feminist Approaches to the Hebrew Bible*, ed. S. Scholz. Oxford: Oxford University Press.

Gimbutas, Marija. 1974. *The Gods and Goddesses of Old Europe, 7000 to 3500 B.C: Myths, Legends and Cult Images*. Berkeley: University of California Press.

Habel, Norman C., ed. 2000. *Reading the Bible from the Perspective of the Earth*. The Earth Bible, 1. Sheffield: Sheffield Academic Press.

———., ed. 2001. *The Earth Story in the Psalms and Prophets*. The Earth Bible, 4. Sheffield: Sheffield Academic Press.

———., ed. 2009. *An Inconvenient Text: Is a Green Reading of the Bible Possible?* Hindmarsh, SA: ATF Press.

———., ed. 2012. *Rainbow of Mysteries: Meeting the Sacred in Nature*. Kelowna, BC: Copper House.

Habel, Norman C., and Vicky Balabanski, eds. 2002. *The Earth Story in the New Testament*. The Earth Bible, 5. London: Sheffield Academic Press.

Habel, Norman C., and Peter Trudinger, eds. 2008. *Exploring Ecological Hermeneutics*. SBL Symposium Series 46. Atlanta: SBL.

Habel, Norman C., and Shirley Wurst, eds. 2000. *The Earth Story in Genesis*. The Earth Bible, 2. Sheffield: Sheffield Academic Press.

———, eds. 2001. *The Earth Story in Wisdom Traditions*. The Earth Bible, 3. Sheffield: Sheffield Academic Press.

Hadley, Judith M. 1997. From Goddess to Literary Construct: The Transformation of Asherah into Hokmah. In *A Feminist Companion to Reading the Bible: Approaches, Methods and Strategies*, ed. A. Brenner and C. Fontaine, 360–399. Sheffield: Sheffield Academic Press.

Hampson, Daphne, and Rosemary Radford Ruether. 1987. Is There a Place for Feminists in a Christian Church? *New Blackfriars* 68 (801, Jan.): 7–24.
Haraway, Donna. 2008. *When Species Meet.* Posthumanities, 3. Minneapolis: University of Minnesota Press.
Hobgood-Oster, Laura. 2001. Wisdom Literature and Ecofeminism. In *The Earth Story in Wisdom Traditions*, ed. N. C. Habel and S. Wurst, 35–47. The Earth Bible, 3. Sheffield: Sheffield Academic Press.
Irigaray, Luce. 1986. *Divine Women.* Trans. S. Muecke. Sydney: Local Consumption.
Kelly, Anthony. 1993. *An Expanding Theology: Faith in a World of Connections.* Sydney: E. J. Dwyer.
Levine, Amy-Jill. 2004a. Response. *Journal of Feminist Studies in Religion* 20 (1): 125–132.
———. 2004b. The Disease of Postcolonial New Testament Studies and the Hermeneutics of Healing. *Journal of Feminist Studies in Religion* 20 (1): 91–99.
Merchant, Carolyn. 1990 [1980]. *The Death of Nature: Women, Ecology and the Scientific Revolution.* Republished with New Preface. San Francisco: Harper and Row.
Pattel-Gray, Anne. 1995. Not Yet Tiddas: An Aboriginal Womanist Critique of Australian Church Feminism. In *Freedom and Entrapment: Women Thinking Theology*, ed. M. Confoy, D. Lee, and J. Nowotny, 165–192. Melbourne: Dove.
———. 1999. The Hard Truth: White Secrets, Black Realities. *Australian Feminist Studies* 14 (30): 259–266.
Plaskow, Judith. 1991. Feminist Anti-Judaism and the Christian God. *Journal of Feminist Studies in Religion* 7 (2, Fall): 99–108.
Plumwood, Val. 1993. *Feminism and the Mastery of Nature.* London: Routledge.
———. 2002. *Environmental Culture: The Ecological Crisis of Reason.* London: Routledge.
Primavesi, Anne. 2000. *Sacred Gaia: Holistic Theology and Earth System Science.* London: Routledge.
———. 2003. *Gaia's Gift: Earth, Ourselves and God after Copernicus.* London: Routledge.
Rigby, Kate. 1998. Woman and Nature Revisited: Ecofeminist Reconfigurations of an Old Association. *Arena Journal* 12: 143–169.
———. 2001. The Goddess Returns: Ecofeminist Reconfigurations of Gender, Nature and the Sacred. In *Feminist Poetics of the Sacred: Creative Suspicions*,

ed. F. D. Glass and L. McCreddin, 23–54. New York: Oxford University Press.
———. 2004. Earth, World, Text: On the (Im)Possibility of Ecopoiesis. *New Literary History* 35: 427–442.
———. 2008. Noah's Ark Revisited: (Counter-)Utopianism and (Eco-) Catastrophe. *Arena Journal* 31: 163–177.
———. 2014. 'Come Forth into the Light of Things': Material Spirit as Negative Poetics. In *Material Spirit: Religion and Literature Intranscendent*, ed. G. C. Stallings, M. Asensi, and C. Good, 111–128. New York: Fordham University Press.
———. 2015. *Dancing with Disaster: Environmental Histories, Narratives, and Ethics for Perilous Times*. Charlottesville: University of Virginia Press.
Rose, Deborah Bird, Thom van Dooren, Matthew Chrulew, Stuart Cooke, Matthew Kearnes, and Emily O'Gorman. 2012. Thinking through the Environment: Unsettling the Humanities. *Environmental Humanities* 1 (1): 1–5. http://environmentalhumanities.org/arch/vol1/EH1.1.pdf
Routley, Richard, and Val Routley. 1973. *The Fight for the Forests: The Takeover of Australian Forests for Pine, Wood Chips, and Intensive Forestry*. Canberra: Australian Research School of Social Sciences.
Ruether, Rosemary Radford. 1983. *Sexism and God-Talk: Toward a Feminist Theology*. London: SCM Press.
———. 1992. *Gaia and God: An Ecofeminist Theology of Earth Healing*. San Francisco: HarperCollins.
Schüssler Fiorenza, Elisabeth. 1983. *In Memory of Her: A Feminist Theological Reconstruction of Christian Origins*. London: SCM Press.
———. 1992. *But She Said: Feminist Practices of Biblical Interpretation*. Boston: Beacon Press.
Sittler, Joseph. 2004 [1964]. *The Care of the Earth*. Adapted from *The Care of the Earth and Other University Sermons*. Minneapolis: Fortress Press.
Spivak, Gayatri Chakravorty. 1993. French Feminism Revisited. In *Outside in the Teaching Machine*, 141–172. London: Routledge.
Stanton, Elizabeth Cady. 1897. Letter from Elizabeth Cady Stanton. In *Memoriam: Frederick Douglass*, ed. H. Douglass, 44–45. Philadelphia: J.C. Yorston & Co. http://antislavery.eserver.org/legacies/frederick-douglass-elizabeth-cady-stanton/frederick-douglass-elizabeth-cady-stanton-pdf
———. ed. 1993. *The Woman's Bible*. 2 vols. First published New York: European Publishing C., 1895–1898. Republished with a forward by Maureen Fitzgerald. Boston: Northeastern University Press.

Starhawk. 1989 [1979]. *The Spiral Dance: A Rebirth of the Ancient Religion of the Great Goddess. 10th Anniversary Edition*. San Francisco: HarperSanFrancisco.

Trible, Phyllis. 1978. *God and the Rhetoric of Sexuality*. Philadelphia: Fortress Press.

———. 1984. *Texts of Terror*. London: SCM Press.

Wainwright, Elaine M. 1991a. *Towards a Feminist Critical Reading of the Gospel According to Matthew*. Berlin: De Gruyter.

———. 1991b. A Metaphorical Walk through Scripture in an Ecological Age. *Pacifica* 4 (3): 273–294.

———. 1995. A Voice from the Margin: Reading Matthew 15:21–28 in an Australian Feminist Key. In *Reading from this Place, Vol. 2: Social Location and Biblical Interpretation in Global Perspective*, ed. F. S. Segovia and M. A. Tolbert, 132–154. Minneapolis: Augsburg Fortress.

———. 1997. Rachel Weeping for Her Children: Intertextuality and the Biblical Testaments—A Feminist Approach. In *A Feminist Companion to Reading the Bible: Approaches, Methods and Strategies*, ed. A. Brenner and C. Fontaine, 452–469. Sheffield: Sheffield Academic Press.

———. 1998. *Shall We Look for Another? A Feminist Rereading of the Matthean Jesus*. Maryknoll: Orbis Books.

———. 2000. A Transformative Struggle towards the Divine Dream: An Ecofeminist Reading of Matthew 11. In *Reading the Bible from the Perspective of the Earth*, ed. N. C. Habel, 162–173. The Earth Bible, 1. Sheffield: Sheffield Academic Press.

———. 2002. Which Intertext? A Response to an Ecojustice Challenge: Is Earth Valued in John 1? In *The Earth Story in the New Testament*, ed. N. Habel and V. Balabanski, 83–88. The Earth Bible, 5. London: Sheffield Academic Press.

———. 2003. The Pouring Out of Healing Ointment: Rereading Mark 14:3–9. In *Toward a New Heaven and a New Earth: Essays in Honor of Elisabeth Schüssler Fiorenza*, ed. F. F. Sergovia, 157–178. Maryknoll: Orbis Books.

———. 2005. Looking Both Ways or in Multiple Directions: Doing/teaching Theology in Context into the Twenty-first Century. *Pacifica* 18 (2): 123–140.

———. 2006. *Women Healing/Healing Women: The Genderization of Healing in Early Christianity*. London: Equinox.

———. 2008a. Healing Ointment/Healing Bodies: Gift and Identification in an Ecofeminist Reading of Mark 14:3–9. In *Exploring Ecological Hermeneutics*, ed. N. C. Habel and P. Trudinger, 131–140. SBL Symposium Series, 46. Atlanta: SBL.

———. 2008b. Unbound Hair and Ointmented Feet: An Ecofeminist Reading of Luke 7.36–50. In *Exchanges of Grace. Festschrift for Ann Loades*, ed. N. K. Watson and S. Burns, 178–189. London: SCM Press.
———. 2009. The Book of the Genealogy: How Shall We Read It? *Concilium* 3. In *Eco-Theology* (themed issue), ed. E. Wainwright, L. C. Susin and F. Wilfred, 13–23.
———. 2012. Images, Words, Stories: Exploring Their Transformative Power in Reading Biblical Texts Ecologically. *Biblical Interpretation* 20: 280–304.
———. 2013. Of Borders, Bread, Dogs and Demons: Reading Matthew 15.21–28 Ecologically. In *Where the Wild Ox Roams: Biblical Essays in Honour of Norman C. Habel*, ed. A. H. Cadwallader and P. Trudinger, 114–128. Sheffield: Sheffield Phoenix Press.
———. 2015. Queer[y]ing the Sermon on the Mount. In *Sexuality, Ideology and the Bible Antipodean Engagements*, ed. R. J. Myles and C. Blyth, 115–131. Sheffield: Sheffield Phoenix.
———. 2016. *Habitat, Human, and Holy: An Eco-Rhetorical Reading of the Gospel of Matthew*. The Earth Bible Commentary Series. Sheffield: Sheffield Phoenix.
White, Lynn. 1967. The Historical Roots of our Ecological Crisis. *Science* 155 (3767, Mar.): 1203–1207.

Anne Elvey is an adjunct research fellow, School of Languages, Literatures, Cultures and Linguistics, Monash University; Honorary Research Associate, Trinity College Theological School; and a member of the Centre for Research in Religion and Social Policy, University of Divinity, Melbourne. Her publications include: *Ecological Aspects of War: Engagements with Biblical Texts* (coedited with Keith Dyer and Deborah Guess, 2017). Anne is managing editor of *Plumwood Mountain: An Australian Journal of Ecopoetry and Ecopoetics*.

13

Australian Women in Mining: Still a Harsh Reality

Maryse Helbert

From the gold rush in the 1850s to hydraulic fracturing which began in the mid-2000s, Australian economic growth has been heavily dependent on its capacity to dig and extract natural resources for the world market. While the Australian mining industry has produced social and economic benefits, it has also had negative impacts upon sections of Australian society. In this chapter, I apply a materialist ecofeminist critique as a means of showing how the gender gap supports increased mining and the distribution of risks and benefits of the mining industry is due to capitalist patriarchy. I show how capitalist accumulation in mining areas impacts upon and intersects with inequalities of class, gender, ethnicity, race and location. The locations I consider in this chapter are the mining communities of the Bowen Basin and the Century Mine in Queensland as well as the Pilbara, and Pembleton communities in Western Australia. The capitalist patriarchal structures of the mining industry (and its

M. Helbert (✉)
University of Melbourne, Melbourne, VIC, Australia

L. Stevens et al. (eds.), *Feminist Ecologies*,
https://doi.org/10.1007/978-3-319-64385-4_13

spillover into mining communities) are mechanisms that maintain women and Indigenous people in subservient roles of 'unproductive' labour that sustains the male-dominated productive economic system. Emancipatory changes in mining communities might begin by connecting different oppressive structures, a task that materialist ecofeminism is well-placed to perform.

Materialist ecofeminism posits that social reproduction, care work, household chores such as food preparation and subsistence economy are fundamental elements of life needs and survival. However, this work is considered 'unproductive' within a capitalist and patriarchal system which undervalues women's work by associating women's activities with the activities of nature (Salleh 1993: 226; Mies 1998: 37–38). While women's work is being undervalued through the gendered division of labour and the split between the public and private sphere, men's economic activities are given greater priority (Mellor 1997: 130). Male production can then focus on the accumulation of capital while domestic labour is largely left to women (Salleh 1993: 226). This has adverse implications for mining communities.

The Two-Person Career and Women's Subservient Role

There is a rich and complex body of research about the relationship between women and mining industries globally. This research shows that in developing and developed countries alike, there is a deficit of representation of women in the large-scale mining industry.[1] In a study in the 1970s of the oil industry in Calgary, Canada, sociologist John Douglas House noted that 'the oilmen of Calgary include no women, no radicals, no Jews, no native Indians, no Inuit, no East Indians, no blacks, and no French Canadians'. Things have hardly improved 40 years later. In 2008, the Women's rights movement in Ghana demanded better job opportunities in the emerging oil industry, but this was rejected by the Resource Minister who stated that jobs in the oil industry were for men (GhanaWeb 2010). In Norway in 2010, Hege Marie Norheim, Vice President for research and business development of StatoilHydro (Norway's national oil company), points out that 'the most important jobs in the oil industry, the business jobs, are still given mainly to men' (World and Feltus 2010:

72–73). Hence, worldwide, research shows there is a deficit of representation of women in the mining industry. If a larger proposition might be that women reject the environmental destruction brought by mining, as Indigenous women activists have done, it is apparent from the research here that women's unpaid labour often supports mining.

In the Australian context, the poor representation of women in the mining industry is well researched. A 2016 report on workplace gender equality shows that women made up just 15% of the mining workforce (Australia Gender Equality Scorecard 2016: 11). This is further confirmed by the poor representation of women as managers and technical experts within the industry (Australia Gender Equality Scorecard 2016: 11). Social scientist Sanjay Sharma studied the work distribution in the Bowen Basin region, a coal mining region in Queensland. In this region, in 2006, the labour force participation was as follows: 67.3% of men worked full time, while 29.4% of women were employed full time (Sharma 2010: 208; ABS 2007). Nearly three fifths of women in mining towns were either not in the labour force or were employed part time-while a majority of males were employed in the mining industry (47.8%) or in the construction industry (10.9%) (Sharma 2010: 209; ABS 2007). However, only 9.5% of women were employed in the mining industry and 3.5% in the construction industry (Sharma 2010: 210; ABS 2007).

When women are working in the Australian mining industry, the average pay gap against them is 15.8% (Australia Gender Equality Scorecard 2016: 6). Criminologist Kerry Carrington shows in Pembleton, a mining community in the remote region of Armstrong, Western Australia, a greater proportion of males than females received high incomes compared with elsewhere in Australia (Carrington et al. 2010: 400).

Human geographer Robyn Mayes explains that the poor representation of women in the mining industry conceals other more insidious forms of patriarchal dominance (Lozeva and Marinova 2010: 181; Mayes 2014: 122). Women working in the industry have further confirmed its masculinist culture. In 2002, social scientists Joan Eveline and Michael Booth researched the working environment of the Emsite, a remote mine operation in Western Australia's Pilbara region. They showed that women employed in the mining industry had to face overt and persistent sexism and sexual harassment as part of their everyday work life (Eveline and Booth 2002: 149). The women who were interviewed believed that the sexist culture of the mining industry was a manifestation of the men's resistance to their employment (Eveline and Booth 2002: 149). In a qualitative and quantitative survey conducted in the Pilbara mining region of Western Australia in 2015, social scientist Bobana Kljajevic wanted to understand the causes behind the lower number of women in the industry. She showed that while some women were not deterred from working in the mining industry, they were quite aware of dominant patriarchal organizational cultures in the industry (Kljajevic 2015: 144). According to the women interviewed, the patriarchal culture was most acutely illustrated by some male senior managers hiring men rather than women (Kljajevic 2015: 145). The women interviewed also denounced the negative work culture towards them and the lack of enforcement of non-discriminatory policies (Kljajevic 2015: 146). Women identified masculinist values as the cause of their lack of career advancement opportunities (Kljajevic 2015: 146). Preferential treatment of men in the mining workforce in the Pilbara region accounts for the underrepresentation of women in that mining industry.

The discrimination against women in mining industries throughout Australia still exists despite decades of policies and initiatives attempting to boost the number of women employed. Women's groups, government and mining corporations have been actively trying to address the underrepresentation of women in the mining industry for decades (Mayes and Pini 2010). One particular milestone was the equal opportunity movement and the Equal Employment Opportunities Act passed in 1987 which aimed to provide a legal basis to counteract such discrimination against women (Mayes 2014: 127). The business media described the policies and initiatives implemented to boost women number in the mining industry as a 'feminine revolution', particularly if women could reach a critical mass large enough to influence the oil industry's working culture (Mayes and Pini 2010: 238). In the media this 'feminine revolution' depicted men's dominance in the industry as an historical rather than contemporary phenomenon (Mayes and Pini 2010: 238). Indeed, Mayes and Pini show how the media's claim of a 'feminine revolution' in the mining industry suggests that women have feminine advantages due to a range of naturally occurring attributes. These 'feminine advantages' act, according to the media, as a counterbalance to the masculinist structures of the mining industry (Mayes and Pini 2010: 234). But, as Mayes and Pini conclude, women who are admitted into the industry tend to try to blend in rather than foreground their gender. For instance, a female mine manager interviewed said: 'I'm not one of those women's champions. I just get on with the job and I happen to be a woman' (Mayes and Pini 2010: 239). Despite efforts to improve the relation between women and the mining industry, it remains dominated by men. Rather than being a sign of improvement, the belief that a 'feminine revolution' has already taking place in the sector reinforces the masculinist values of the mining industry.

Wage levels in the mining industry are approximately double the average weekly earnings in the retail trade and industries (ACIL Consulting 2002). Women are more likely to be employed in the non-mining industry sector, but these jobs are less likely to be as well paid as in the mining industry. In the mining towns of the Bowen Basin region, for example, women were mostly employed in the retail trade, education and training, accommodation and food services and health care and social assistance (Sharma 2010: 210; ABS 2007). These jobs reflect the traditional patriarchal characterization of women as primarily fit to be 'carers' (Salleh 1993: 227). As the masculinist cultures of the mining industry spill over into the mining communities, the sexual division of labour is perpetuated, pushing women further towards the unpaid work of sustaining the mining communities.

The concept of a two-person career, developed by social scientist Hanna Papanek, shows how reproduction, care work, household and subsistence economy are often hidden within the private sphere (Papanek cited in Rhodes 2003: 150). As Papanek notes, the two-person career is a combination of informal and formal work, in which women are confined to the unpaid work of the household and men to the well-paid jobs in the mining industry (Papanek cited in Rhodes 2003: 150). This sexual division of labour means women are predominantly left to perform the supportive and nurturing work that is considered 'unproductive' within capitalist societies. In the context of mining communities, the wives of male workers are often expected to cook and host dinner parties, a well-entrenched popular activity in these isolated regions that have few other forms of Western cultural entertainment. These dinner parties contribute to forging and consolidating business relationships in the mining communities and can also contribute to elevating the personal status of a mining engineer (Papanek cited in Rhodes 2003: 152). The sexual division of labour is perceived by mining companies 'as a firm foundation for successful mining operations' (Papanek cited in Rhodes 2003: 152). The masculinist values of the mining companies are thus aimed at deterring women from entering the industry and instead encourage women to perform labour that sustains the economic activities of the male-dominated institutions.

The unpaid labour that women undertake in mining communities becomes more problematic still when we take into account the economic and social impacts of mining industries upon the local economies. Sharma shows that in the remote mining communities of Australia, certain policies that reinforce the containment of women in the sexual reproductive sphere sustain gender inequality within families and communities (Sharma 2010: 202). Economist John Rolfe shows how the processes of the operation of mines have a social and economic impact on the mining boom in the Bowen Basin in Central Queensland (Rolfe et al. 2007). In the Bowen Basin communities, the rapid expansion of the coal mining sector brought positive outcomes such as job creation in mining and construction (Rolfe et al. 2007: 135). However, the rapid expansion of coal mining also caused the decline of other non-mining sectors such as the service industry, retail trade and tourism sectors (Rolfe et al. 2007).

The rapid expansion that a mining boom generates distorts the local economy of other sectors. The expectation of high-paid jobs puts pressure on the wages in other employment sectors in the region. Higher wages mean higher costs and sectors such as tourism and services may not have the ability to support those extra costs, jeopardizing their existence. Additionally, the non-mining industry can find it difficult to recruit workers due to shortages of labour. As it was shown in the Bowen Basin mining communities, the high incomes that the mining sector offers attract workers, making it more difficult for other sectors to keep up (Rolfe et al. 2007). In the same way, in the energy hub community of Gladstone, in Queensland, where gas is converted to LNG (Liquefied Natural Gas), the high wages offered by the mining companies have created a shortage of labour in the nurse and police force as these workers have sought employment in the higher-paid mining industry (Mitchell-Wittington 2017: 4).

Furthermore, in an extensive review of literature on the well-being of women in mining towns in Australia, Sharma suggests that the remoteness of the mining communities—which often have extreme climatic conditions—pushes women to feel isolated from their friends and family. This leads to women's greater social and economic dependence on their male partners. Hence dependent married women show more vulnerability to mental illness, particularly due to the demands of domestic labour (Sharma 2010: 212). In remote mining communities, women bear the emotional cost of the particular capitalist patriarchal settings in which women primarily do the unpaid work of sustaining the male-dominated economic infrastructure.

Mining Communities and Family Violence

Criminologists Kerry Carrington, Alison McIntosh and John Scott conducted a study into family violence in rural communities in the three largest Australian states—Western Australia, New South Wales and Queensland (Carrington et al. 2010). These communities had experienced rapid community and economic growth due to a mining boom. Their study showed that the combination of the influx of high-paid workers, mostly men, the masculinist culture of the mining jobs and the remoteness of the mining communities increased the levels of family and nonfamily violence and crime and hence the insecurity of the community (Carrington et al. 2010). Traditional explanations of increased violence in rural communities have correlated with times of economic decline and rising unemployment (Carrington et al. 2010: 395). Research around mining communities throughout the world highlights the ways in which mining industry activities exacerbate existing social problems, such as alcoholism, gambling, trafficking and forced labour (Anderson 1998; Haller et al. 2007: 395; Watts 2004: 3).

In Australia, there is a link between rising social disorder and alcohol consumption, and the strength of the link increases with the level of geographic remoteness (Carrington et al. 2010: 401). In the mining community of Pemberton and the Armstrong Mine, located in southwest Western Australia, four work camps have liquor licences and regular private bus services run between the camps and the popular drinking venues (Carrington et al. 2010: 401). This shows how the community of Pemberton service the mining workers by providing facilities for mining workers to socialize. There is also plenty of anecdotal evidence pointing to a link between alcohol consumption and workplace status. A young male explained that in the community: 'Everyone drinks to get drunk… the more drunk you are the more cool points you get' (Carrington et al. 2010: 395). Research conducted by Carrington et al., shows that resource boom communities have, on average, higher rates of violence, debilitating injuries, motor vehicle accidents and suicides in comparison to metropolitan areas (Carrington et al. 2010: 395).

Sexism and the harassment of women in the workplace are further exclusionary mechanisms that make it difficult for women to work in the 'male-controlled institutions' of the mining industry (Salleh 1997: 14). The particular patriarchal settings of mining communities that service its workers have increased levels of violence, making such sites even more hostile for women. Carrington et al. note that popular drinking venues are still very much a male domain. The 'public masculinity' of binge drinking in pubs in the Pemberton mining communities acts as a deterrent for women to go to the pub as 'it is too scary to go there' (Carrington et al. 2010: 401). Violence in public spaces further excludes women from public places, further confining them to the privacy of the home (Carrington et al. 2010). In homes in the Pilbara mining communities the crime rate for domestic assault is far higher than the state average (Gately et al. 2016).

The use of violence to control women is further highlighted by high levels of skimpies and prostitution in these regions. Skimpies are barmaids who serve food and drinks in their underwear, or in some cases work topless, and are found in most rural mining towns in Australia. In Kalgoorlie, skimpies and brothels are viewed by the local community as inevitable due to the notion that 'men have an uncontrollable heterosexuality' (Pini et al. 2013: 173). The belief in 'natural sexual urges' is not a new phenomenon and has led the local communities and the mining companies themselves to discreetly organize prostitution in mining communities by accepting brothels or by organizing fly-in prostitutes (Scott 2013). The local mining community believes that these 'uncontrollable natural sexual urges' need to be released in order to avoid sexual and other kinds of violence (Pini and Mayes 2014: 432; Pini et al. 2013: 173). According to a local in Kalgoorlie, 'if we did not have them (the prostitutes), we would have a lot more rapes and murders' (Pini et al. 2013: 173). Skimpies in Kalgoorlie talk about the horrific 'violent experiences of abuse and harassment they are subjected to' (Pini et al. 2013: 173–174).

The use of skimpies to service the mining workers is further highlighted by their containment to particular designated areas of the community. Indeed, in the Kalgoorlie mining community, for instance, while the community believe that skimpies and prostitutes should be used to service the sexual needs of the mining workers, they also believe skimpies and prostitutes should be policed by 'the behaviour and practices of residents' (Pini et al. 2013: 174). In Kalgoorlie, the skimpies are confined to the 'Skimpie house', a decrepit building with minimum facilities. One justification for their isolation offered by the community is that it is ostensibly for their well-being, for fear that they will be 'potential recipients of violence' (Pini et al. 2013: 174). Yet, their spatial containment is reflected socially in the way skimpies are ignored by the general community and by clients when they venture into the streets outside their workplace. The social exclusion of skimpies and prostitutes in mining communities like Kalgoorlie has to be understood as an insidious and effective social mechanism that contains women and confines them to subservient roles.

In addition to gender issues, Australian Indigenous communities suffer at the hands of an industry which prioritizes the needs of mining workers in mining communities. Indigenous anthropologist and geographer Marcia Langton highlights how Indigenous communities living close to particular mines suffer from what is known as 'the resource curse' (Langton 2010). The concept of the resource curse came out of development economics (theoretically grounded in neoliberalism) and describes the present relations between natural resources endowment and economic growth (Langton 2010; Langton and Mazel 2015: 35–37; Auty 1994b). Economists of the development economics school view a nation's natural resources as an asset for the country's economy (Auty 1994a: 12). However, economist Richard Auty has shown that some developing countries such as Venezuela, Nigeria and Angola do not transform the revenue streams coming out of the extraction of rich and plentiful natural resources into benefits for the whole nation. To illustrate the resource curse in Australia, Langton shows that there is a 'city-bush divide' in her research about Karratha and Roebourne in the southern Pilbara region of Western Australia (Langton 2010). Karratha is a port and dormitory town on the coast of the southern Pilbara regions of Western Australia. Roebourne is an Indigenous community half an hour's drive inland from Karratha. Karratha has amenities such as a motel and shopping centres to service the miners and their families. However, Roebourne is in the bush, 'old and dusty, showing signs of years of neglect' (Langton 2010). The community of Roebourne is what Val Plumwood would call the 'unconsidered background' (Plumwood 2002: 104) to the 'civilization' of Karratha.

The city-bush divide is further illustrated by the underrepresentation of Indigenous people in the mining industry. This underrepresentation is still prevalent despite relations between Indigenous people, the mining industry and the federal government improving over the last 40 years. Several milestones have contributed to this improvement. First, the Mabo vs. Queensland High Court recognition of native title in Australia, which reversed the 'longstanding fiction of terra nullius' (Langton and Mazel 2015: 39), recognizing the traditional rights of Indigenous Australians to their land and waters in common law. The High Court decision was followed by the Native Title Act (NTA) in 1993 that established procedures to deal with the development of natural resources projects. In some cases, the NTA gives native title-holders the right to be notified and consulted about the development of natural resources projects.

However, as anthropologist Deborah Bird Rose showed in 'Women and Land Claims', while there have been improvements in the relations between Indigenous communities and the mining industry, these improvements have developed within particular patriarchal settings. Land claim processes were prepared mostly by men who were largely disinterested in representing the rights of Aboriginal women (Rose 1995: 15). Indeed, Rose shows that the relationships between the government, the mining companies and the Indigenous communities to negotiate land rights largely excluded women and their right to make claims over particular areas of land (1995: 16). In 1998 the NTA was amended to include the introduction of Indigenous land use agreements. This amendment opened the door to negotiations in matters of access to land, resources and infrastructure, but also environmental management, compensation, employment and training opportunities for Indigenous communities impacted by the mining industry (Langton and Mazel 2015: 40–43).

Despite these improvements, in a study of Indigenous communities in the Pilbara mining area, human geographers John Taylor and Ben Scambary show that the engagement of Indigenous people within the mining industry is very recent and the change of economic status for Indigenous people is still limited. They cite a complex set of reasons for this, including Indigenous dependence on government and the limited capacity of Indigenous communities living close to the mines to organize themselves to take advantage of the mining boom (Taylor and Scambary 2006: 1). Overall, and despite improvement, there is still a big gap between the employment rate of the general population and the employment rate of the Indigenous communities, especially Indigenous women in the mining industry.

In a study of Indigenous communities living in remote Western Australia around mining areas, social scientist Alfred Dockery shows that the mining areas 'did not display higher than average Indigenous participation and employment rates' (2014: 83). Dockery concludes that between 2006 and 2011, none of the empirical tests revealed any improvement in Indigenous unemployment rates relating to mining activities (2014: 83). Similarly, social scientist Joni Parmenter notes that Indigenous women at the Century Mine of northwest Queensland are suffering from a double discrimination: sexism and racism (Parmenter 2008: 7). In the 2006 census, it was shown that only 0.4% of the total mining workforce was made up of Indigenous women. The jobs that Indigenous women had in the mining industry were predominantly as cleaners and kitchen hands (2008: 7). The challenge for Aboriginal communities to have their land rights recognized and to be fully integrated in the development processes of the mining industry reflects a long history of denial of their presence. Their lands were appropriated in the same process that considered Indigenous activities inessential to capitalist patriarchy. The denial and neglect of Indigenous women are further indicative of a patriarchal capitalist system that privileges profit above all else.

Historically and today, however, Indigenous communities have been at the forefront of resistance to mining projects. For instance, the Jabiluka Mine in Kakadu National Park in the Northern Territory showed how the Indigenous people of Mirrar, the traditional owners of the mine site, local environmental movements and international organizations worked successfully to resist the mining of uranium. The resistance against uranium mining attracted international attention when the two community leaders, Yvonne Margarula and Jacqui Katona, were arrested for trespassing when protesting against the mine. The two women were awarded the US-based Goldman Prize for their environmental protection work. This award was an encouragement for the Gundjeihmi Aboriginal Corporation (GAC), to which the two women belong, to get the Kakadu National Park on the UN World Heritage (UN WH) 'endangered' list. The GAC and the Jabiluka Action Group (JAG) used legal and international appeals to the UN WH committee to show that mining Jabiluka was illegal (Hintjens 2000: 379). The federal government used what JAG has called 'diplomatic blackmail' towards the UN WH (Hintjens 2000: 379). Despite strong resistance, the mine opened in 2000. Women's resistance to mining is grounded in their recognition of a connection between different oppressive structures. Patriarchal structures affect Indigenous people, particularly women and ignore their mediation with nature. They bear the overwhelming cost of the white male dominated economic infrastructure. All oppressive structures need to be addressed for their emancipation.

Conclusion: Towards an Ecofeminist Ethics in Mining Communities

The strength of ecofeminism is its capacity to shed light on the connections between different oppressive structures and to recognize potential emancipatory possibilities. Ecofeminist ethics argues for the creation of different, non-hierarchical and integrative ways of viewing the world (Plumwood 2002: 168). According to Plumwood and Salleh, deconstruction of the patriarchal system and its attitudes towards the environment and women will involve revaluing the contribution and moral worth of both women and nonhuman nature (Salleh 2006: 12, 14, 2001: 3; Plumwood 2002: 168). In mining communities such as Bowen Basin, the Century Mine, Kalgoorlie, Pilbara, and Pembleton, there is a clear gender gap in the distribution of the economic gains of mining. Using an ecofeminist approach I have shown that this gap is due to the particular gendered settings that create circumstances where women are encouraged and, in some cases, coerced into labour that sustains the social infrastructure that privileges men's economic advancement. This impedes women's choices, opportunities, agency and autonomy while their unpaid work is too often usurped for corporate profit. It may also impact on anti-mining activism.

Notes

1. Large-scale mining denotes the scale of the mining activities using large engine to dig, blast and extract the natural resources. It is a highly mechanized process usually planned by a large corporation. Small-scale mining such as artisanal mining use rudimentary techniques. It can be legal or illegal, formal or informal. There is a larger number of women working for small-scale mining (Hinton et al. 2003).

Bibliography

ABS. 2007. *Australian Bureau of Statistics 2006. Catalogue No. 2901.0—Census dictionary, 2006*. Canberra: ABS.

ACIL. 2002. *Queensland's Mining Industries: Creating Wealth for the Community, the State and the Nation*. Brisbane: ACIL Consulting.

Anderson, Kym. 1998. Are Resource-Abundant Economies Disadvantaged? *Australian Journal of Agricultural and Resource Economics* 42 (1): 1–23.

Auty, Richard M. 1994a. Industrial-Policy Reform in 6 Large Newly Industrializing Countries—The Resource Curse Thesis. *World Development* 22 (1): 11–26.

———. 1994b. The Resource Curse Thesis—Minerals in Bolivian Development, 1970–90. *Singapore Journal of Tropical Geography* 15 (2): 95–111.

Carrington, Kerry, Alison McIntosh, and John Scott. 2010. Globalization, Frontier Masculinities and Violence: Booze, Blokes, Brawls. *British Journal of Criminology* 50 (3): 393–412.

Dockery, Alfred Michael. 2014. The Mining Boom and Indigenous Labour Market Outcomes. In *Resource Curse or Cure?: On the Sustainability of Development in Western Australia*, ed. M. Brueckner, A. Durey, R. Mayes, and C. Pforr. Heidelberg: Springer.

Eveline, Joan, and Michael Booth. 2002. Gender and Sexuality in Discourses of Managerial Control: The Case of Women Miners. *Gender, Work and Organization* 9 (5): 556–578.

Gately, Natalie, Suzanne Ellis, and Robyn Morris. 2016. Drug Use Monitoring in Australia: An Expansion into the Pilbara. *Trends and Issues in Crime and Criminal Justice* 504: 1–14.

Haller, Tobias, Annja Blochlinger, Markus John, Esther Marthaler, and Sabine Ziegler, eds. 2007. *Fossils Fuels, Oil Companies and Indigenous Peoples: Strategies of Multinational Oil Companies, States, and Ethnic Minorities: Impact on Environment, Livelihoods and Cultural Change*. London: Transaction Publishers.

Hintjens, Helen. 2000. Environmental Direct Action in Australia: The Case of Jabiluka Mine. *Community Development Journal* 35 (4): 377–390.

Hinton, Jennifer J., Marcello M. Veiga, and Christian Beinhoff. 2003. Women and Artisanal Mining: Gender Roles and the Road Ahead. In *The Socio Economic Impacts of Artisanal and Small-Scale Mining in Developing Countries*, ed. G. Hilson. Lisse, The Netherlands: Taylor and Francis.

Kljajevic, Bobana. 2015. An Investigation into the Underrepresentation of Women in the Pilbara Mining Region of Western Australia. PhD thesis, Department of Social Sciences and International Studies, Curtin University.

Langton, Marcia. 2010. The Resource Curse. *Griffith Review*, 28, online. Accessed 11 March 2017. https://griffithreview.com/articles/the-resource-curse/

Langton, Marcia, and Odette Mazel. 2015. Poverty in the Midst of Plenty: Aboriginal People, the 'Resource Curse' and Australia's Mining Boom. *Journal of Energy and Natural Resources Law* 26 (1): 31–65.

Lozeva, Silvia, and Dora Marinova. 2010. Negotiating Gender: Experience from Western Australian Mining Industry. *Journal of Economic and Social Policy* 13 (2): 177–209.

Maijita, Ali. 2010. Women Unsuited for Work in Oil Sector—Asaga. In *Ghana Oil and Gas*. Accra: African Media Empire Ltd.

Mayes, Robyn. 2014. Gendered Dimensions of Resource Extraction: The Place of Women. In *Resource Curse or Cure?: On the Sustainability of Development in Western Australia*, ed. M. Brueckner, A. Durey, R. Mayes, and C. Pforr. Heidelberg: Springer.

Mayes, Robyn, and Barbara Pini. 2010. The 'Feminine Revolution in Mining': A Critique. *Australian Geographer* 41 (2): 233–245.

Mellor, Mary. 1997. Women, Nature, and the Social construction of 'Economic Man'. *Ecological Economics* 20 (3): 129–140.

Mies, Maria. 1998. *Patriarchy and Accumulation on a World Scale: Women in the International Division of Labour*. London: Zed Books.

Mitchell-Wittington, Amy. 2017. Pipe Dreams. *The Age*.

Parmenter, Joni. 2008. Considering the Experience of Indigenous Women in the Australian Mining Industry. *Gender and Mining Conference*. Canberra: Australian National University.

Pini, Barbara, and Robyn Mayes. 2014. Performing Rural Masculinities: A Case Study of Diggers and Dealers. In *Masculinities and Place*, ed. A. Gorman-Murray. Farnham: Ashgate Publishing.

Pini, Barbara, Robyn Mayes, and Kate Boyer. 2013. 'Scary' Heterosexualities in a Rural Australian Mining Town. *Journal of Rural Studies* 32: 168–176.

Plumwood, Val. 2002. *Environmental Culture: The Ecological Crisis of Reason*. London: Routledge.

Rhodes, Linda. 2003. Partners on the Periphery: Personal Ambiguity and Unpaid Labour in the Australian Mining Industry. *Journal of Australian Studies* 27 (76): 147–158.

Rose, Deborah Bird. 1995. Women and Land Claims. Native Title Research Unit, Australian Institute of Aboriginal and Torres Strait Islander Studies: 1–8.

Rolfe, John, Bob Miles, Stewart Lockie, and Galina Ivanova. 2007. Lessons from the Social and Economic Impacts of the Mining Boom in the Bowen Basin 2004–2007. *Australasian Journal of Regional Studies* 13 (2): 134–153.

Salleh, Ariel. 1993. Class, Race, and Gender Discourse in the Ecofeminism/Deep Ecology Debate. *Environmental Ethics* 15 (3): 225–244.
———. 1997. *Ecofeminist as Politics: Nature, Marx and the Postmodern*. London: Zed Books.
———. 2001. The Capitalist Division of Labour and its Meta-Industrial Class. *Tasa 2001 Conference*. Sydney: University of Sydney.
———. 2006. Moving to an Embodied Materialism. *Capitalism Nature Socialism* 16 (2): 9–14.
Scott, John. 2013. Fly-In Crime? *Arena Magazine* 124: 44–46.
Sharma, Sanjay. 2010. The Impact of Mining on Women: Lessons from the Coal Mining Bowen Basin of Queensland, Australia. *Impact Assessment and Project Appraisal* 28 (3): 201–215.
Taylor, John, and Ben Scambary. 2006. *Indigenous People and the Pilbara Mining Boom: A Baseline for Regional Participation*. CAEPR research monograph/Centre for Aboriginal Economic Policy Research, Australian National University: No. 25. Canberra: Australian National University E Press.
Watts, Michael. 2004. Antinomies of Community: Some thoughts on Geography, Resources and Empire. *Transactions of the Institute of British Geographers* 29 (2): 195–216.
Workplace Gender Equality Agency. 2016. *Australia's Gender Equality Scorecard: Key findings from the Workplace Gender Equality Agency's 2015–2016 Reporting Data*. Canberra: Workplace Gender Equality Agency.
World Petroleum Council and Anne Feltus. 2010. *Women in Energy: Closing the Gender Gap*. London: World Petroleum Council.

Maryse Helbert has been an advocate for, and researcher on, women's participation in politics and decision-making for over a decade. After completing a Master's thesis on the comparative strengths and weaknesses of the French and Finnish political systems in encouraging and increasing women's political participation (University of Rennes 2, France, and Abo Akademi, Finland), she became actively involved in the movement to institute the so-called Parity Law in France (1999–2000). Her PhD, *Women in the Oil Zones: A Feminist Analysis of Oil Depletion, Poverty, Conflict and Environmental Degradation*, exposes the sidelining of women from decision-making processes at local, national and international levels on issues of energy security, and challenges the paradigms of corporate social responsibility in the extractive industries. Maryse has recently worked as a gender-based violence expert for the United Nations' Sustainable Development Goals and as a researcher with the Association of Women in Development. She currently teaches global politics and political economy at the University of Melbourne, Australia.

14

'In the Interest of All Mankind': Women and the Environmental Protection of Antarctica

Emma Shortis

This chapter highlights the major contribution of women activists working against mining in Antarctica during the 1980s, and reinstates them in the historical record. Despite a growing recognition of the need to include women and gender in studies of the Antarctic, the history of the continent remains overwhelmingly dominated by accounts of men. This chapter first points out how women do, in fact, have a long history on the continent, as companions, explorers, scientists, and activists, and then describes the feminist histories of women that reinvigorate scholarship on the Antarctic. It points out, however, that this work does not include studies of women environmental activists, who were central to efforts to protect the continent from environmental destruction. A greater understanding of women's deep engagement with the continent also illuminates Antarctica's larger role in the international environmental movement. Finally, this chapter analyses how gendered perceptions of the continent

E. Shortis (✉)
The University of Melbourne, Melbourne, VIC, Australia

L. Stevens et al. (eds.), *Feminist Ecologies*,
https://doi.org/10.1007/978-3-319-64385-4_14

as 'white' and 'pure' rebound on female presence, and the complicated interplay of such gendered perceptions with legitimate desires to protect the Antarctic environment.

The 1959 Antarctic Treaty mandates that Antarctica 'shall continue forever to be used exclusively for peaceful purposes and shall not become the scene or object of international discord'. This, the treaty makers assert, is 'in the interest of all mankind' (Antarctic Treaty Secretariat 1959). Scholars and activists have generally assumed that 'all mankind' can be safely assumed to mean 'all humankind'. Historically, however, Antarctica has been an almost exclusive preserve of 'mankind' only. Womankind, at least until very recently, has been almost entirely absent both from the continent and our study of it (Dodds 2009). As the preserve of 'mankind', the Antarctic has been understood and constructed in gendered terms, as a 'virgin' continent, and a space in which to test and demonstrate a particularly masculine (and Western) form of heroism. This designation of the Antarctic as a masculine space marginalizes the role of women on the continent, and, in the words of Klaus Dodds, has significant implications for 'how and with what consequences the polar continent has been settled and studied' (2009: 505).

Women, however, have a long history on the continent, initially as companions to their conquering husbands, and later as scientists and activists in their own right. Gender, furthermore, is central to understanding desires to first conquer, and then protect, the Antarctic. Women activists were in fact central to efforts to protect the continent, in what became one of the most significant political developments in Antarctic history: the defeat of a nearly completed international agreement that would have allowed mining to begin on the continent, in favour of a new agreement that banned mining and committed signatories to comprehensive environmental protection. This was an extraordinary achievement for both the environmental movement and the individuals and states involved. Existing histories of this agreement, however, do not address the central role of environmental activists and their campaigns in its defeat. The fact that some of the central players within this movement were women is rarely mentioned. These women activists, however, were one of the major driving forces behind the successful environmental protection of an entire continent. How gendered understandings of this continent as 'fragile', 'pure', and 'pristine' played into desires to protect it from environmental destruction are also central to understanding the effective public appeal of the campaign. An understanding of the central role of women in Antarctic history and politics, and broader feminist and gender-based approaches, is thus crucial to understanding the past, present, and future of the continent.

Feminist Approaches

Feminist approaches are only just beginning to find a voice amongst studies of the Antarctic. After being among the first Australian women to visit the continent in an official capacity, Elizabeth Chipman published *Women on the Ice* in 1986. Chipman's book was the first in-depth look at the history of women in the Antarctic. While the book was certainly groundbreaking, Chipman cautions that she 'felt no obvious discrimination' when she visited Antarctica (1986: 6). The book thus traces the history of women's involvement in Antarctica, but fails to interrogate the reasons behind the erstwhile exclusion of women, or to consider the popular characterization of Antarctica as a masculine space. It was Lisa Bloom's *Gender on Ice: American Ideologies of Polar Expeditions* in 1993 that was the first to consider the deeper ramifications of gender for the history, and study, of the Antarctic. These two pioneering works were followed by others, including Esther Rothblum, Jacqueline Weinstock and Jessica Morris' *Women in the Antarctic* in 1998, and Robin Burns' *Just Tell them I Survived! Women in Antarctica* in 2001. Both Burns and Rothblum had themselves visited the Antarctic, and both books focus on tracing the history of women, and their experiences, on the continent. Several significant contributions have also been published in a 2009 edition of the journal, *Signs: Journal of Women in Culture and Society*.

While relatively small, this significant body of work on women, gender, and the Antarctic has the potential to make a significant impact on polar studies. Collectively, this work asserts that historically, Antarctica has been seen as a 'virgin continent', ripe for exploration, conquest, and, most importantly, as a stage on which to demonstrate virile, masculine heroism. In the ironic, but accurate, words of Christy Collis: 'Ban women from half a continent, and pretty quickly that half continent becomes a fantasy world for masculinity' (2009: 514). While Collis is referring specifically to the Australian Antarctic Territory, her words are reflective of the history of the continent as a whole.

Collis is not alone in pointing out the absence of women in Antarctic history. Popular perceptions, and even many gender-based histories of the Antarctic, tend to regard women as almost entirely absent from the continent until at least the 1960s. Rosner asserts that this is mostly true, as women have not been prominent as explorers, conquerors, or researchers of the Antarctic (2009: 491). As Dodds argues, women's role 'was at best marginal' (2009: 507). Even those 'marginal' women, furthermore, have rarely had their roles or work acknowledged, either by the men they were with or by mainstream scholarship.

Travel Companions to the Conquerors

Women, however, do have a significant history in the Antarctic, and tracing this history was Chipman's great original contribution. As she points out, there is evidence of women visiting the sub-Antarctic as early as 1773, and women were present well into the nineteenth century, mostly as wives and companions to captains of whaling and sealing expeditions (1986). It was not until 1935, though, that a Norwegian woman, Caroline Mikkelsen, would actually set foot on the continent itself (Chipman 1986). Like most of the women before her, Mikkelsen had accompanied her husband on a whaling expedition, but she was the first to disembark and place her feet on land (Chipman 1986: 75; see also Lewander 2009: 92). It was not until more than a decade later that two women, Edith Ronne and Jennie Darlington, members of a private expedition, would spend a winter on the continent (Burns 2001: 15). Like the women before them, neither was there in a professional capacity (Chipman 1986: 113).

The Ronne expedition marked a significant milestone in women's presence on the Antarctic continent. However, the fact that women had now proved that they could survive a winter on the continent did not increase women's presence in the Antarctic. After Ronne, as Chipman observes, national governments increasingly controlled access to the Antarctic, and government-led and funded expeditions were closed to women (1986: 67). As Dodds puts it, 'women were considered unsuitable for such an environment' (2009: 506–507) and would have to fight for their right to visit the continent they were studying.

The first four Australian women to participate in a government expedition went to the sub-Antarctic Macquarie Islands in the summer of 1959–1960 (Burns 2005: 2). Isobel Bennet, one of those women, later reflected that 'We were invaders in a man's realm and were regarded with some suspicion. We had been warned that on our behaviour rested the future of our sex...' (quoted in Chipman 1986: 44). Further, while two male-led and manned expeditions led by the Dutch explorer Roald Amundsen and British explorer Robert Falcon Scott had arrived within weeks of each other at the South Pole in 1911, it was not until the summer of 1969–1970, almost a decade after Bennet's first visit, that women arrived at the South Pole as part of an American expedition for the very first time (Chipman 1986: 95). It was, then, not until the beginning of the 1970s that women scientists were able to participate in continental expeditions (Dodds 2009: 508). Australian women visited Casey Station, on the continent itself, for the first time in the summer of 1975–1976, during International Women's Year (IWY) (Burns 2001: 21). Despite this small increase in women's presence in the Antarctic, the continent was still not entirely open to women. American women were not allowed to spend winter on the continent until 1974, while British women had to wait until the 1990s (Dodds 2009: 508), nine years after the first Australian woman, Louise Holliday, in 1981 (Burns 2005: 3). This milestone, somewhat unsurprisingly, still did not mean that all barriers had been broken down (Burns 2001: 20). Chipman, for example, argues that the two American women who wintered on the continent in 1974—one of whom was Chief Scientist—were chosen because they were old and unattractive, claiming that '[t]he women were not likely to be sexually provocative' (111–113). While Chipman acknowledges that the women were 'the best in their fields', even she subscribes to the view that it was women's sexuality—not their brains—that was paramount (113; see also Burns 2001: 23). It is this focus on sexuality that overwhelmingly dominates histories of Antarctica. Outside the specific feminist and women's histories outlined above, women's historic role on the continent is always overshadowed by heroic tales of male competition and conquest.

The fact that women do have a history in the Antarctic, however, should by now be clear. Like men, women have been active in the sub-Antarctic and on the continent itself since at least the 1770s. Women have travelled to the continent as companions to invaders and explorers, and later worked there as scientists in their own right. Today, however, women still represent a significant minority on Antarctic research stations, despite the fact that, as Chipman wrote in 1986, '[w]omen have proved they can do the job in Antarctica as elsewhere' (7).

Women Activists and the Campaign for a 'World Park' Antarctica

The little scholarship that is available on women's historic role in the Antarctic does not address the role of women activists; as outlined above, the focus is overwhelmingly on women as companions, private expeditioners, or state-sanctioned scientists. This reflects not only the dearth of gender-based scholarship on Antarctica but also the general lack of research into the role of non-state activists on the continent. While the role of non-state activists and non-governmental organizations (NGOs) in, for example, the international campaign to ban whaling has been studied comprehensively (Keck and Sikkink 1998: 127), their role in the Antarctic campaign of 1989–1991 is barely understood (Shortis 2015; Clark 2013). Environmental NGOs' ability to disseminate and promote the idea that Antarctica should be protected from mining, however, is central to understanding how the parties to the Antarctic Treaty were convinced that six years of negotiation over minerals exploitation should be abandoned in favour of comprehensive environmental protection.

The Convention on the Regulation of Antarctic Mineral Resource Activities, which would have allowed mining to begin on the Antarctic continent, was adopted by the parties to the Antarctic Treaty in June 1988 (CRAMRA 1988). Environmental organizations such as Greenpeace, the World Wide Fund for Nature, the Wilderness Society, the International

Fund for Animal Welfare, and the Australian Conservation Foundation, united under the Antarctic and Southern Ocean Coalition (ASOC), had begun campaigning for the environmental protection of Antarctica nearly a decade before the Minerals Convention was signed (Shortis 2015; ASOC 2013). ASOC was the leading non-governmental coalition advocating for Antarctica, representing over 200 member organizations in 35 countries (Antarctica Project 1989). After the Convention was signed, the international campaign against Antarctic mining sought to convince parties to the Treaty not to ratify the agreement.

In Australia, that campaign was spearheaded by Lyn Goldsworthy, the coordinator of the Australian branch of ASOC. Australia, as a claimant to 42% of Antarctic territory, plays a significant role in Antarctic politics, and so in the 1980s Goldsworthy was well placed to influence both Australian Antarctic policy and broader international developments (Haward and Griffiths 2011: 102). In her role in the Australian branch of ASOC, Goldsworthy executed a brilliant campaign aimed at stopping the Australian government from ratifying the Convention, which it had spent more than six years negotiating. Goldsworthy and her colleagues worked closely with government policy makers, at times cleverly exploiting interdepartmental tensions. While continuing this highly sophisticated behind-the-scenes lobbying, ASOC Australia, under Goldsworthy's direction, simultaneously used more traditional techniques like newsletters, petitions, and advertisements, which all helped to generate substantial interest from both the media and the public. The strong public response to the campaign fed directly into the eventual decision by both the Australian and French governments to oppose the advent of mining in Antarctica (Shortis 2015; Clark 2013: 170–171). In 1989, the Australian Prime Minister, Bob Hawke, came to the conclusion that Australia should *not* ratify the Minerals Convention, claiming that 'it was inconceivable that we should put at risk the one remaining pristine continent' (Hawke 1994: 467–468). The Australian government's sudden refusal to ratify the Minerals Convention, given that the Antarctic Treaty System operates under a consensus rule, meant that the Minerals Agreement could not be adopted. Existing histories of the Australian decision against ratification focus overwhelmingly on the role of government, and a more recent spat between the former Prime Minister and his former Treasurer, Paul Keating, around who decided what first (Haward and Griffiths 2011: 246–247; Bramston 2015). While the significant role of such tensions, and broader concerns about the sovereignty of the Australian Antarctic Territory, have been acknowledged by the existing scholarship, the extent and nature of Goldsworthy's role in Australian decision-making has not. The significant role played by the coordinator of ASOC Australia in this momentous decision, and the subsequent Australian campaign to replace the Minerals Convention with an environmental protection agreement, is rarely acknowledged. Goldsworthy, however, deserves much of the credit both for the original decision and later the Australian government's successful negotiating role in the lead up to the 1991 Madrid Protocol for Antarctica. Goldsworthy's work has however been officially acknowledged when she was made a Member of the Order of Australia for her contributions to conservation.

Women also played a significant role in the international campaign against Antarctic mining, which worked closely with national campaigns like the one led by Goldsworthy. Kelly Rigg was the coordinator of the Greenpeace International Antarctic Campaign from 1986 to 1990. Rigg was one of the driving forces behind the internationalization of the campaign. She coordinated and organized several Greenpeace expeditions to Antarctica, which were crucial for both attracting media and public attention to the issue, and also in elevating Greenpeace into a respected voice in Antarctic politics (Clark 2013: 164–165). Women were on each one of these expeditions, increasing the presence of women activists and scientists on the ice itself. Rigg was also crucial in enlisting the French celebrity filmmaker and adventurer Jacques-Yves Cousteau for the campaign. In 1989, Cousteau started a petition against the Convention in France, and in less than a year had 1.5 million signatures (Committee on Merchant, Marine and Fisheries 1990: 5). An American version, organized jointly by the Cousteau Society and ASOC, also gained 1.5 million signatures in less than a year (AP 1989).

Official players rarely acknowledge the role played by international environmental organizations in generating this significant and influential public response to the prospect of mining in Antarctica. Ex-Prime Minister Hawke, for example, notes the 'growing anxiety around the world on global environmental issues', but his retelling of the story only acknowledges the work of the Australian and French governments and Cousteau (Hawke 1994: 468). Hawke does not acknowledge (indeed, does not even mention) the work of Goldsworthy, Rigg, Greenpeace, or ASOC in helping to generate and focus this 'growing anxiety' on Antarctica. As Hawke acknowledges, Cousteau most certainly played a central role in the campaign against the Minerals Convention, spearheadeding an intensive lobbying campaign in the United States, which, under President George H W Bush, was in favour of Antarctic mining. He generated substantial media coverage, organized public petitions, which were signed by millions of people, met with congressmen and senators, appeared before Congress and its various Subcommittees on multiple occasions, and even took six children—one from each continent—on a highly publicized awareness-raising expedition to the Antarctic. Cousteau, along with Greenpeace International and Greenpeace USA (where the Antarctic campaign was also headed by a woman), was instrumental in generating the political pressure that would eventually sway the Bush administration into supporting an environmental protection agreement for Antarctica (for more on Cousteau's significant role, see Shortis 2015). During this campaign, Cousteau worked closely with Greenpeace and ASOC. Together, environmental organizations, Cousteau, and the French and Australian governments ensured that the Mining Convention was defeated and replaced by a comprehensive environmental protection agreement. Women activists like Rigg and Goldsworthy played an essential role in devising, directing, and executing this sophisticated and successful campaign, which resulted in the pre-emptive protection of an entire continent.

A 'Virgin' Land

These women activists were undoubtedly working in a space, both real and imagined, dominated by men. As New Zealand activist Cath Wallace observed, women working on the campaign in the 1980s got the clear message that 'especially young women who aren't even scientists' shouldn't have a say in Antarctic politics (interview with author 2014). She felt that 'many of the officials and scientists [she] dealt with were incredibly patronizing', and that she and her colleagues faced 'extraordinary hostility'. Women like Wallace, Rigg, and Goldsworthy were in the double bind of challenging from the outside a state-sponsored, previously settled agreement in a closed diplomatic environment. They were also, simultaneously, focused on a particular environment both dominated by men, and imagined as an exclusively masculine preserve.

During the campaign for a 'World Park' Antarctica in the 1980s, despite the significant presence of women activists, gendered understandings of the continent persisted. As late as 1990, Jacques Cousteau was referring to the Antarctic as 'this virgin land' (*Tulsa World* 1990). Virginity, of course, is associated with purity—a characterization of Antarctica that was deliberately perpetuated during the 'World Park' Antarctica campaign. Antarctica was, as Cousteau described it, 'the last unspoiled area of our planet' (*Tulsa World* 1989, 1990). To Australian environmentalists, 'Antarctica is the ultimate wilderness, the last and greatest' (Suter 1979: 9). Politicians and the media took up this message as well. For Hawke 'it was inconceivable that we should put at risk the one remaining pristine continent' (Hawke 1994: 467–468). This characterization of the 'last pristine wilderness' is not limited to decades' past. Even today, in serious academic publications, Antarctica is still labelled 'pristine' without much concern or qualification (Haward and Griffiths 2011; Leane 2007: 262). These understandings of Antarctica as 'pure', 'pristine', 'unspoiled', and so on align with broader narratives in nature protection which value purity and tend to feminize nature. In the 1980s, environmentalists like Cousteau, Rigg, Goldsworthy, and Wallace cleverly integrated this representation of Antarctica as fragile and vulnerable into the contemporary emotional language of environmentalism, carving out a role for the ice as an emotional symbol for the environmental movement (Fay 2011: 293). Partly as a result of this campaign, and growing knowledge of the scientific value of a 'pristine' Antarctic, the continent became, in historian Tom Griffiths' words, 'the key to the future of humanity' (2010: 28). That 'key', though, was implicitly feminine. As Victoria Rosner has so clearly outlined, '[i]n common references to the poles as pure, pristine, or untouched, we hear echoes of the old talk about the seventh, virgin continent, so chilly and remote yet so sought after by men' (2009: 493).

Historically, Antarctica has been understood primarily as a site for male competition and conquest. Women have been excluded from the space both physically and emotionally—physically by governments and expeditions restricting their presence or banning them entirely, and emotionally, as too weak to cope with the demands of an isolated and hostile environment. Femininity was, and arguably still is, allowed only as a framework for understanding the fragility and purity of a place that must be protected by men. As Bloom observes in *Gender on Ice*, 'These last spaces on earth, which still remained invisible and therefore inscrutable, excited a consuming passion on the part of white men of various Western countries to "conquer" and make "visible" these sites' (1993: 3). These 'lusts', as Lewander describes them, were certainly not considered appropriate for women, whose weakness and fragility had no place on a continent that needed controlling and conquering (2009: 93). Rosner labels this the 'grand heroic tradition', which 'defines the polar regions as all-male spaces of bonding, conquest and noble suffering' (2009: 490). Women, in this context, cannot possibly be the 'heroes', because, in effect, it is femininity itself that is being conquered (Lewander 2009: 93).

This designation of the Antarctic as an exclusively male preserve has meant that demonstrations of male 'heroism' have imbued popular histories, and perceptions, of the Antarctic (Rosner 2009: 491). Women activists do not fit into this framework. According to Bloom, both the Arctic and the Antarctic thus 'occupy a peculiar position' in the development of British and American nationalism (1993: 3). Certainly, for Britain, and the Australian and New Zealand colonies, the history of Antarctic conquest and exploration is closely tied to imperial narratives. In national imaginations, the Antarctic became a place where men went to test, and demonstrate, their imperial masculinity, and through that, the masculine superiority of the nation/empire itself (Rosner 2009: 490; Dodds 2009: 505; Bloom 1993: 6). It should not be surprising that women could not fit into this context, and at least partly explains the unwillingness to include women in Antarctic expeditions and the hostility faced by women activists. Women, should they be allowed to participate, would shatter the mythic masculine space of the Antarctic. As American Rear Admiral George Dufek said in 1959, 'I felt the men themselves didn't want the women there. It was a pioneering job. I think the presence of women would wreck the illusion of the frontiersmen—the illusion of being a hero' (quoted in Chipman 1986: 87).

Conclusions

The indefinite ban on mineral resource exploitation in the Antarctic, and the internationally agreed upon environmental protection of the entire continent, represents an almost unparalleled achievement in the history of international environmentalism. In the words of one environmental campaigner, it was 'arguably Greenpeace's biggest victory ever' (Kelly Rigg, interview with author 2014). Surprisingly, however, there is very little existing research into this extraordinary feat. Given the relative lack of success stories in international environmental negotiations, especially more recently, this is a significant oversight. Women activists were central to the success of the 'World Park' Antarctica campaign of the 1980s. In 1991, that campaign culminated in one of the most significant international environmental agreements in existence today, an indefinite ban on Antarctic mining. Women activists, however, were one of the major driving forces behind the unprecedented success of this campaign, and an understanding of the central role of women in Antarctic history and politics, and broader feminist and gender-based approaches, is crucial to understanding the past, present, and future of the continent. At the centre of international political questions of resource management and exploitation, conquest, sovereignty, and the environment, the Antarctic continues and will continue to play a significant role in international relations. As the Antarctic comes under pressure from tourism, climate change, and political tensions, the ramifications of gender-based approaches should not be underestimated. The Antarctic's history as a gendered space will continue to influence its future and goes some way to explaining the resonance of campaigns to preserve and protect its 'fragile' environment. Deeper understandings of women's historic role in Antarctic activism, and the importance of gender to both the continent and environmental activism more broadly, might offer alternative understandings of, and approaches to, the future of an environment under threat.

Bibliography

Antarctic and Southern Ocean Coalition Secretariat. 2013. History of Antarctic and Southern Ocean Coalition. Accessed 6 June 2016. http://www.asoc.org/about/history

Antarctic Treaty Secretariat. 1959. *The Antarctic Treaty*. Accessed 1 December 2016. http://www.ats.aq/e/ats.htm

———. 1988. *Convention on the Regulation of Antarctic Mineral Resource Activities (CRAMRA)*. Accessed 6 June 2016. http://www.ats.aq/documents/recatt/Att311_e.pdf

AP. 1989. Antarctica Meeting Opens with Calls for Preservation. *Tulsa World*, October 10.

Bloom, Lisa. 1993. *Gender on Ice: American Ideologies of Polar Expeditions*. Minneapolis: University of Minnesota Press.

Bloom, Lisa, Elena Glasburg, and Laura Kay. 2008. Introduction. *The Scholar and Feminist Online: Gender on Ice* 7 (1, Fall). Accessed 16 August 2013. sfonline.barnard.edu/ice/print_intro.htm

Bramston, Troy. 2015. Cabinet Papers 1988–1989: Hot Under Collar at Who Saved Antarctica. *The Australian*. Accessed 1 January 2015. http://www.theaustralian.com.au/in-depth/cabinet-papers/cabinet-papers-1988-89/cabinet-papers-19881989-hot-under-collar-at-who-saved-antarctica/news-story/7039ad7f176f4c4292b0b695230b5411

Burns, Robin. 2001. *Just Tell Them I Survived!: Women in Antarctica*. Crows Nest: Allen & Unwin.

———. 2005. Women in Antarctica: Sharing This Life-Changing Experience. *4th Annual Phillip Law Lecture*. Hobart. June 18.

Clark, Margaret. 2013. The Antarctic Environmental Protocol: NGOs in the Protection of Antarctica. In *Environmental NGOs in World Politics*, ed. M. Finger and T. Princen, 160–185. London: Taylor and Francis.

Chipman, Elizabeth. 1986. *Women on the Ice: A History of Women in the Far South*. Melbourne: Melbourne University Press.

Collis, Christy. 2009. The Australian Antarctic Territory: A Man's World? *Signs: Journal of Women in Culture and Society* 34 (3): 514–519.

Committee on Merchant, Marine and Fisheries, House of Representatives; Subcommittee on Oceanography and Great Lakes. 1990. Antarctica Briefing with Jacques-Yves Cousteau: Hearing before the Subcommittee on Oceanography and Great Lakes of the Committee on Merchant Marine and Fisheries, House of Representatives, One Hundred First Congress, Second

Session; on a Briefing on Mr. Cousteau's Recent Expedition to Antarctica in the Hope of Increasing Environmental Knowledge and Awareness, and Preserving the Continent for Future Generations, May 2, 1990. Washington: U.S. Government Printing Office.

Dodds, Klaus. 2009. Settling and Unsettling Antarctica. *Signs: Journal of Women in Culture and Society* 34 (3): 505–509.

Edwards, Kerry, and Robyn Graham, eds. 1994. *Gender on Ice: Proceedings of a Conference on Women in Antarctica held in Hobart, Tasmania, under auspices of the Australian Antarctic Foundation.* Canberra: Australian Government Publishing Service.

Elliot, Lorraine M. 1994. *International Environmental Politics: Protecting the Antarctic.* New York: St Martin's Press.

Fay, Jennifer. 2011. Antarctica and Siegfried Kracauer's Cold Love. *Discourse* 33 (3, Fall): 291–321.

George Bush Presidential Library, College Station, TX, Letter, The Antarctica Project to George Bush, 23 June 1989, document range: 207988 to 253707, 9 S SF CO, CO001-02, Countries, Antarctica—Arctic, WHORM, Subject File—General, ID# 049914, Bush Presidential Records.

Greenpeace International, M. V. Greenpeace Crew List, Antarctic Expedition 1985/86. 1985. p. 1 Folder: Greenpeace Expedition 1985–86, MS 9432, Box 1, Records of the Antarctic Campaign, National Library of Australia (NLA), Canberra.

———. 1989a. Aquino Rules out Seeking Re-Election Series: WORLD DIGEST. *Tampa Bay Times*, October 20, 22A.

———. 1989b. Sunday's People. *The Record*, October 22, np.

———. 1990. Convicted Lawmaker to Resign. *Tulsa World*, January 3, np.

Griffiths, Tom. 2010. A Humanist on Thin Ice. *Griffith Review* 29 (Spring): 22–37.

Haward, Marcus, and Tom Griffiths, eds. 2011. *Australia and the Antarctic Treaty System: 50 Years of Influence.* Sydney: UNSW Press.

Hawke, Robert J. 1994. *The Hawke Memoirs.* London: William Heinemann.

Joyner, Christopher C. 1998. *Governing the Frozen Commons: The Antarctic Regime and Environmental Protection.* Columbia: University of South Carolina Press.

Keck, Margaret E., and Kathryn Sikkink. 1998. Environmental Advocacy Networks. In *Activists Beyond Borders: Advocacy Networks in International Politics*, 121–163. Ithaca: Cornell University Press.

Leane, Elizabeth. 2007. 'A Place of Ideals in Conflict': Images of Antarctica in Australian Literature. In *The Littoral Zone: Australian Contexts and Their Writers*, ed. C.A. Cranston and R. Zeller, 261–289. Amsterdam: Rodopi.

Lewander, Lisbeth. 2009. Women and Civilization on Ice. In *Cold Matters: Cultural Perceptions of Snow, Ice and Cold*, ed. H. Hansson and C. Norberg, 89–103. Umea: Umea University and the Royal Skyttean Society.

Lowthian, Mary. 1987. Review: Women in Antarctica. *Polar Record* 23: 603.

Milton, Kay. 2002. *Loving Nature: Towards an Ecology of Emotion*. London: Routledge.

Rigg, Kelly. 2014. Interview with author, March 10.

Rosner, Victoria. 2009. Gender and Polar Studies: Mapping the Terrain. *Signs: Journal of Women in Culture and Society* 34 (4, Spring): 489–494.

Rothblum, Esther D., Jacqueline S. Weinstock, and Jessica Morris, eds. 1998. *Women in the Antarctic*. New York: Harrington Park Press.

Shortis, Emma. 2015. 'Who Can Resist this Guy?' Jacques Cousteau, Celebrity Diplomacy, and the Environmental Protection of the Antarctic. *Australian Journal of Politics and History* 61 (3): 363–378.

Suter, Keith D. 1979. *Antarctica: World Law and the Last Wilderness*. Sydney: Friends of the Earth.

Wallace, Cath. 2014. Interview with author, February 18.

Emma Shortis is a PhD Candidate in the School of Historical and Philosophical Studies at the University of Melbourne, researching the history of the environmental protection of the Antarctic. In 2014, she was the recipient of a National Library of Australia Summer Scholarship, and in 2013, the recipient of a Society for Historians of American Foreign Relations Global Scholars Grant. In 2017, she was awarded a Fox International Fellowship at Yale University for the academic year 2017–2018.

Index[1]

[1] Note: Page number followed by 'n' refers to notes.

L. Stevens et al. (eds.), *Feminist Ecologies*, https://doi.org/10.1007/978-3-319-64385-4

MIX
Papier aus verantwortungsvollen Quellen
Paper from responsible sources
FSC® C105338

Printed by Books on Demand, Germany